THE BATTLES AT PATTERSON AND FORT BENTON

It was "A War Within A War"
Civil War Commission of Missouri Pamphlet

BOB FRAKES

© 2024 by Bob Frakes. All rights reserved.

Words Matter Publishing
P.O. Box 1190
Decatur, IL 62525
www.wordsmatterpublishing.com

No part of this publication may be reproduced, stored in a retrieval system, or transmitted in any way by any means—electronic, mechanical, photocopy, recording, or otherwise—without the prior permission of the copyright holder, except as provided by USA copyright law.

ISBN 13: 978-1-962467-56-8

Library of Congress Catalog Card Number: 2024950363

DEDICATION

I would like to dedicate this book to David Hagler. I had a number of discussions with David over the years, and he created the chapter on Fort Benton and the sketch of that fort used in this book. David passed away a few years ago, and I appreciate his support and contributions.

ACKNOWLEDGMENTS

I want to also thank my wife, parents, brother, grandparents, aunts and uncles and my many cousins. They are all a part of this book and are noted in the final chapter.

Jimmy Sexton and the River Hills Traveler Magazine for usage permission.

The Civil War Centennial Commission of Missouri for their excellent and often forgotten pamphlet I saved all these years. It was originally given to State Representative Harry Raiffie, who gave it to my Aunt Anna Belle who lived on Delmar and who gave it to me all those years ago. It pays at times to be a pack-rat.

The people at Piedmont City Hall have been ready to help as well as the Wayne County Historical Society.

The archivist at Fort Davidson who was very helpful.

Carole Goggin for the editing of the "Big Picture" chapter.

Scott House for his help & input.

My thanks also go out to Tammy at Words Matter Publishing. And the people of Missouri who have been so helpful with both my "Missouri" books.

TABLE OF CONTENTS

Forward xi

Chapter One – The Big Picture 1
    The Colonies 2
    Confederation and Constitution 3
    An Industrial Revolution Promotes Cotton 4
    Alien & Sedition Acts 5
    Louisiana Purchase 5
    War of 1812 5
    Missouri Compromise 6
    Jackson vs. Calhoun 7
    War with Mexico 7
    Rise of the Abolitionists 8
    Uncle Tom's Cabin 10
    The Kansas-Nebraska Act 10
    The Dred Scott Decision 12
    Final Straws 12

Chapter Two – Missouri 15
    A Divided State in a Divided Nation 15
    The Guerrilla Years: A War of Revenge 23
    The Dying, Desperate Gamble That Lost 29

Chapter Three – Fort Benton fell twice to the South in the Un-Civil War – by David Hagler. 33
    The Fort 34
    Base of Operations. 35
    Confederate Fort. 36

Chapter Four – History of the Battle of Pilot Knob 43
    The Desperate Summer. 44
    The Ragged Assembly 48
    The Fateful March. 48
    The Silence 50
    The Storm. 51
    The Great Escape 52

Chapter Five – Patterson Cemetery 57

Chapter Six – Artifacts from the Farm - The Bayonet 59
    Smoothbore Molds 61
    The Minie' Ball. 63
    3 Inch Archer Bolt. 65

Chapter Seven – Topography and Drone Shots 67
    Fort Benton. 67
    Stoney Battery. 68

Chapter Eight – "Epilogue - The Farm" 71
    The People 71
    The House. 77
    The Spring 79
    The Orchard. 80
    Granary, Shop, and Barn 81
    Rings Creek (Ring's Creek) 82

Snakes Alive!..84
Animal Farm..84
Rings Creek School..87
"The Farm" from Fort Benton.................................88

Picture Notes & Credits......................................**89**

Bibliography/References.....................................**91**

Notes..**93**

References/Suggested Reading................................**95**

Index..**97**

FORWARD

Question – Which states had the largest number of Civil War battles? Well, the answer is Virginia, Missouri, Tennessee, OR, Virginia, Tennessee, and Missouri - depending on the book you read. In fact, the Civil War is often advertised as an "out east" thing, but you have: the Missouri Compromise, Bleeding Kansas (the actual start of the Civil War), the Dred Scott decision (the old courthouse in St. Louis before going to Washington), Wilson's Creek (one of the major battles of the Civil War), and many other events.

In some ways, history seems to have been in my blood. My Dad, Harold Frakes, was born and raised just a few miles from the Lincoln cabin near Gentryville, Indiana. Lincoln's mother, Nancy Hanks Lincoln, is buried there. Dad's Grandmother could actually recall stories of a young Abe Lincoln about town. Abe Lincoln slept here; bed & breakfast sites abound.

On my Mother's side, she was born and raised on Rings Creek on a farm on whose site several "Battles of Patterson" (Patterson, Mo. & Wayne County) had been partly fought with Fort Benton on the hill to the north. When plowing the fields in the spring, Grandpa would overturn artifacts now and then. On the far back part of the farm, I recall three stones in a row, about the size of basketballs. The story goes these stones mark the graves of a pioneer family that got into a dispute with some local Native Americans.

Perhaps it was natural that I would teach American History for decades in the public school system at Waltonville High School, Waltonville, IL.

The Battles at Patterson and Fort Benton

Following the publication of my book on Missouri Lookout Towers, I decided it was time to perhaps get some things down on paper. On the following pages, you will find "The Big Picture," the Civil War in Missouri, the Battles of Patterson and Fort Benton, some drone shots of the area, Stoney Battery, Fort Davidson, the cemetery at Patterson, and more.

I also could not resist a "post script" section on "The Farm" with all the stories and my relatives. I always thought as a youngster, everyone had a farm to go to. Only when I got older did I realize how lucky I had been.

The book is a collection of sorts. I sought input from several sources, and I appreciate their contributions and granted usage.

If you have questions or comments, my phone # is 618-244-1642, and email is frakes2@mvn.net.

Chapter One
THE BIG PICTURE

As one lawmaker observed, slavery was the snake that lay coiled under the table at every deliberation. It had vexed the country since its creation. It seemed to defy solution. It seems compromising moral issues can be like that.

Slavery had been dealt with in many ways. Southern ministers traced slavery's origins back thousands of years. Politically, the Gag Rule forbid its mention in Congress. It was not referred to as slavery by Southern Colonies but "our peculiar institution." There was one attempt after another to reach a compromise. For example, the Missouri Compromise attempted to keep the number of slave and free states equal.

The Civil War would actually be fought for a myriad of reasons. Originally, the Union cause took precedent over slavery. At least, that is what President Lincoln said when he proclaimed, "My paramount object in this struggle is to save the Union, and is not either to save or to destroy slavery. If I could save the Union without freeing any slave, I would do it, and if I could save it by freeing some and leaving others alone, I would do that." After Antietam and the Emancipation Proclamation, the war would be fought for a higher cause as well. There were other ideas afoot. Some wanted only to stop the spread of slavery and were not abolitionists. Jefferson had thought it would take generations for settlers to cross the continent. It happened in the blink of an eye.

Then there were the economic diversities. The North was becoming more industrialized, but the South clung to "king cotton" and the need for slavery and plantations. One of the concepts of modern economy is that it is very hard for a country to exist with two diametrically opposed economic systems.

The war would also be fought by many in the South who did not own slaves but believed their plantation neighbors had that right. When asked why they were fighting by a northerner, many Southerners replied—because you are down here.

How did we get to this point with friend fighting friend and brother fighting brother?

The Colonies

The colonies of Colonial North America were varied by reason of formation and geography (the people, order, soil and crops). These factors would have a big effect on the institution of slavery. In the New England area, farms were small and self-sufficient. The growing season was short. The use of slavery was not profitable. At times, New Englanders disavowed slavery because it was not advantages to them economically, so it was pocket-book easy for them to say, "You should not do that."

In the Middle Colonies, the growing season was longer, and the farms were larger. Indentured servants were, as well as slaves, used in places. Indentured servants were individuals who signed a contract for a period of service in return for their passage to America. These were often the more downtrodden of Europe. Negros who used this device usually had much longer periods of servitude. There were some short-lived attempts to use Native Americans as slave labor, but knowing the area better than the colonists, they simply ran away.

The Southern Colonies had much in common with the Middle Colonies, except indentured servants and later slaves were used to work plantations of indigo, rice and tobacco. Due to the effort needed to remove seeds and the predating of mechanical looms, cotton was not yet "king" and was only grown in central South Carolina. Most Southerners did not own slaves or plantations, but those that did exercised a greater influence over society, politics, and economics.

Throughout the 1600s, the use of indentured servants would shrink, and the use of slaves would increase. The crops were tobacco, rice, and indigo. Some farms in the Middle and Southern Colonies grew regular food crops. With a longer growing season, corn, for example, did very well in the Middle Colonies. Even on the plantations, the rotation of crops was practiced although big money in tobacco caused many to plant that same crop year after year – with destructive consequences to the soil.

Tobacco was such a big hit because John Rolfe had crossed a native Indian crop that would grow but was bitter with a milder English version. In a strange "twist" of history, tobacco was thought to be good for your health, and soon, hogsheads of tobacco tended by indentured servants or slaves filled the docks from Maryland to North Carolina with rice and indigo from North Carolina to Georgia. Cotton would continue to play second fiddle until mechanics, of all things, broke the bottleneck. The explorers had come for God, Gold, and Glory. The colonies produced little to no gold. But by the early 1700s, in the Middle and Southern colonies, it was tobacco, rice, and indigo.

Confederation and Constitution

The Revolutionary War brought the Colonial period to an end. This ushered in our initial and often forgotten first national government. The Continental Congress, out of necessity, had run the affairs of a striving nation during the Revolution. There was now a call to set up something formal. This was accomplished with the Articles of Confederation. Since the national government was so limited, the actual definition of a confederation states that had slavery had no fear of national interference. That does not mean the Articles did not have a pronounced effect on the institution of slavery. The Northwest Ordinance (as in the old Northwest, as in north and west of the Ohio River) stated the area could be carved into no more than five states. Five, it was Ohio, Indiana, Michigan, Illinois, and Wisconsin. The Ordinance also forbad slavery. Having already been demonstrated that slavery was not catching on economically up north, the south saw little to be had in fighting that battle. No effort was made to address to institution as a whole, as the southern states would not have bought in.

When the Articles faltered, the Constitutional Convention would face the same problem. The Convention is at times criticized for not dealing with the slavery issue, but the fact was that slave-holding states would have to buy in, or no document would get signed.

Unlike the Articles, the Constitution would mention slavery in a few ways. Since representation in the House would be based on population, slave-holding states wanted to count slaves as population. Northern states obviously objected. The 3/5 Compromise of 1787 would count 3/5 of the slaves for population numbers. Since the Electoral College granted votes based on Senate and House numbers, this would give slave states an enlarged voice in presidential elections. Also, other provisions stated the slave trade could not be addressed until 1808. This provision would prove of little effect as by 1808, there would be a large domestic supply of slaves. As long as slaves were considered "property," property protections in the Constitution would also help slave states. All told, slave states got much of what they wanted in return for their support.

An Industrial Revolution Promotes Cotton

Toward the end of the 1700s, tobacco was beginning to fade. It had become less fashionable in Europe to use it. Decades of one crop growth had left much of the soil of the "Old South" nutrient-poor and depleted. Cotton growth was stunted by the poor soil, with time-consuming and intensive labor required to remove the seeds. The need was there. Samuel Slater had brought the design for mechanical looms to America in his mind, as England tried to guard the concept and the small rivers of New England were useful to harness water power. The textile industry would help fuel the Industrial Revolution, one of the most influential revolutions in all history. Machines could now crank out woven cotton cloth, with cotton supply being the bottleneck. Eli Whitney would solve the problem. Whitney was the inventor of the interchangeable parts concept. Now, his cotton engine, or "gin," with the turn of a crank, could likewise crank out cotton free of seeds freeing slaves up to attend to other tasks. If only fertile land to the west of the "Old South" could be obtained. Much of the richest black earth in the country was owned by Native American tribes. Cotton was on the verge of becoming "King," bringing with it the expansion of slavery.

Alien & Sedition Acts

These acts passed in 1798, had little to do directly with slavery. But, the reaction to them subtly did. One effect was the Virginia and Kentucky Resolutions. Stated in these were the principles of nullification and interposition, which provided that states were not bound to abide by rules made in Washington, which they felt did not fit the intent of the Constitution. The implication might be subtle, but in the back of some lawmaker's minds was that if they could establish this principle now, any future attempts to stop slavery could simply be declared null and void by a state or states.

Louisiana Purchase

In 1803, the United States purchased from France the large area of "Louisiana". This would be the largest land deal in American History and double the size of the United States. Lost in the purchase was the exact extent of the purchase, the legality of it (was France selling something it actually owned), the fact that many Native Americans already lived there, and the issue of slavery. Of note is the fact that the holdings of what later became the focus of the Kansas-Nebraska Act and a springboard for the Civil War were in its confines. The question of new land and slavery was beginning to heat up.

War of 1812

This "Second American Revolution" had little to do with slavery on the surface. However, "Old Hickory" Andrew Jackson, on his way to win the Battle of New Orleans, had defeated the Creek Indians and forced them to sign a treaty ceding huge tracts of land. The British at New Orleans would be defeated by Jackson's army firing ironically from behind cotton bales. So, Eli Whitney had figured out how to remove the seeds from picked cotton, Sam Slater had figured out how to use mechanical looms to spin textiles, and cotton was no longer confined to the "Old South." The push of Native Americans west would become official government policy. Cotton was becoming "King" with its need for slave labor, and the institution of slavery would begin to wrap itself around every issue.

Missouri Compromise

At the turn of the century, prior to the Louisiana Purchase, the area we call today the State of Missouri wasn't even a part of the United States. After the Purchase, so rapid was the expansion of Americans into the new west that now, in 1820, this new land would be thrust center stage in the debate over slavery and its spread. Although settlement often tended to move east to west in "parallels," Missouri, in fact, had been largely settled by people moving from the South. When Missouri asked for admission to the Union in 1818, North and South sprung to action to try and ensure its admission as a slave state or free state. There were northern attempts to add non-slavery amendments to its admission and southern threats to leave the Union if that was successful. Both sides realized control of the emerging west would allow them to dominate decisions in Congress and the country. Senator J.B. Thomas of Illinois submitted legislation that would become the Missouri Compromise of 1820. It had four major provisions: 1) Missouri would be admitted as a slave state, 2) Maine as a free state (keeping the balance of free and slave the same), 3) the line 36° 30' (the southern border of Missouri) would be drawn west to the Rockies and, 4) slavery would be "forever" (remember that term) prohibited north of that line. This compromise would work for 30 years.

Jackson vs. Calhoun

The next issue came in 1832 with the Tariff of 1832. South Carolina immediately declared it null and void. President Jackson asked for and received a bill authorizing force to back South Carolina down and a reduction in the tariff to placate. South Carolina, noticing no southern state joined them, withdrew its Nullification Ordinance. Both sides claimed victory. In South Carolina the cry was they have lowered the tariff. Jackson pointed to the withdrawal of nullification. Many saw in the affair a pretense; if you can nullify this, then you can nullify any anti-slavery legislation in the future. As the midpoint of the century approached, things would heat up fast as the country tried to deal with slavery and ever more land. The country would stretch to the Pacific in short order.

War with Mexico

The war with Mexico was fought for several reasons, none directly connected to slavery. However, the terms of the treaty ending the war did. With the Mexican Cession, the Gadsden Purchase, the settlement of the

Oregon question, and the annexation of Texas, the United States swept to the Pacific in under a decade. It seemed many issues were tied to slavery. The annexation of Texas had been delayed out of fear the land would become new slave soil. Any attempt to find a permanent solution to the issue seemed to be doomed to failure.

Rise of the Abolitionists

With slavery front page news on a daily basis, the nation became more divided. On one side of the debate, the abolitionist movement arose. Although anti-slavery movements had long existed, the abolitionist movement wanted more than just the freeing of slaves. It called for education and the incorporation of former slaves into society on an equal footing. It started mostly in New England. William Lloyd Garrison founded the anti-slave paper, *The Liberator* (1831-1865). The Underground Railroad began ushering escaped slaves to the north, and the anti-slavery Liberty party was formed, growing strength as the movement grew. In some areas of the south, slaves outnumbered free Southerners 9 to 1. When the Nat

Turner revolt occurred coincidentally with the birth of *The Liberator*, many Southerners moved from describing slavery as "our Peculiar Institution." It was now defended on the following terms: 1) It was common is the Ancient World and supported by scripture, 2) It was essential to the economy of the South, more effective than "wage slavery" practiced in the North, and 3) It would put the slave in a better situation than life in Africa would offer. The chasm between North and South grew wider and wider.

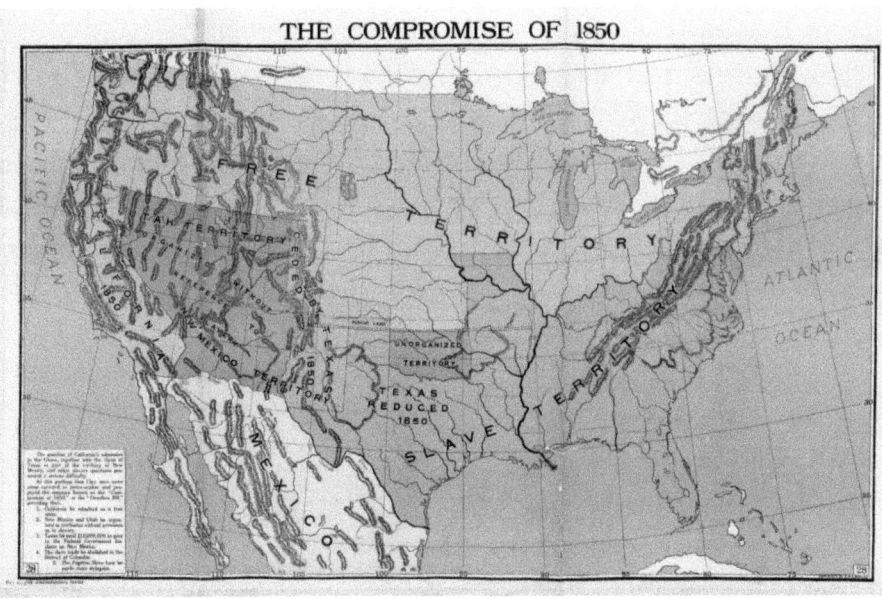

Vast stretches of land to the west now sat unaddressed by a slavery solution. David Wilmot of Pennsylvania proposed that slavery be banned from any territory acquired from Mexico. This would fan the flames and send the leaders in Congress scurrying for a solution. What emerged was the so-called Compromise of 1850. The Compromise was actually a series of proposals seeking a political balance which included: 1) California would be admitted as a free state, 2) the rest of the former Mexican Territory would be organized without restrictions on slavery, 3) the establishment of a western boundary for Texas, 4) the slave trade would not be allowed in D.C, but slavery permitted, and 5) a stronger fugitive slave law which required northerners to return runaway slaves. Many people from different sections of the country objected to parts of the compromise. Northerners especially disliked the Fugitive Slave Law, but others also bought in as they saw this as the chance for the dust to settle on the slavery issue. The Missouri Compromise had lasted 30 years, but the Compromise of 1850 wouldn't.

Uncle Tom's Cabin

Is this the little lady who made this big war? Urban legend has Lincoln stating this upon meeting Harriet Beecher Stowe. Queen Victoria wept as she read the book. The book was *"Uncle Tom's Cabin"* by Harriet Beecher Stowe. The debate over its accuracy in depicting slavery would rage, but many Northerners (who had never observed slavery) saw this as an accurate depiction of the institution. It was a very big log on the fire with only a little more kindling needed for things to go up in flames. A railroad west would provide that spark.

The Kansas-Nebraska Act

There seemed to be a growing rush to bind the nation east and west before it blew apart north and south. Settlers were already pushing west from Missouri and Iowa into unorganized areas. With the settlement of California, there was also a growing demand to build a railroad to the Pacific. It was obvious to Stephen Douglas, Senator from Illinois, that whichever city and state provided the jumping-off point to the west would experience a big financial advantage. It was also obvious to him that Southerners preferred a line through Texas and the New Mexico Territory. A line starting from Chicago would pass through unorganized land. To this end, Douglas proposed the Kansas-Nebraska Act which narrowly passed in May 1854. The Compromise of 1850 had lasted only four years. The Kansas-Nebraska Act said: 1) the unorganized land to the west would be divided into two territories – Kansas and Nebraska. It was assumed one would be free and one slave. 2) The settlers of each territory would decide the slavery issue for themselves – popular sovereignty. Since Kansas had been largely settled by migration from Missouri, free soilers began moving into Kansas to offset this when the vote came. When early efforts at government produced a proslavery legislature, free soilers met and drew up their own government, which President Pierce refused to recognize. During this debate in Congress, Preston Brooks of South Carolina severely beat Charles Sumner of Massachusetts with a cane. Brooks constituents sent more canes, and members of Congress began carrying guns. Violence broke out in Lawrence and along Pottawatomie Creek, where the followers of abolitionist John Brown massacred several proslavery campers. Over the next few months, hundreds would die in what was now called "Bleeding Kansas." Other efforts to admit Kansas as this or that failed, and it would not enter the Union until after the Civil War began. Instead of solving the problem, popular sovereignty was just escalating it. In effect, the Civil War had begun.

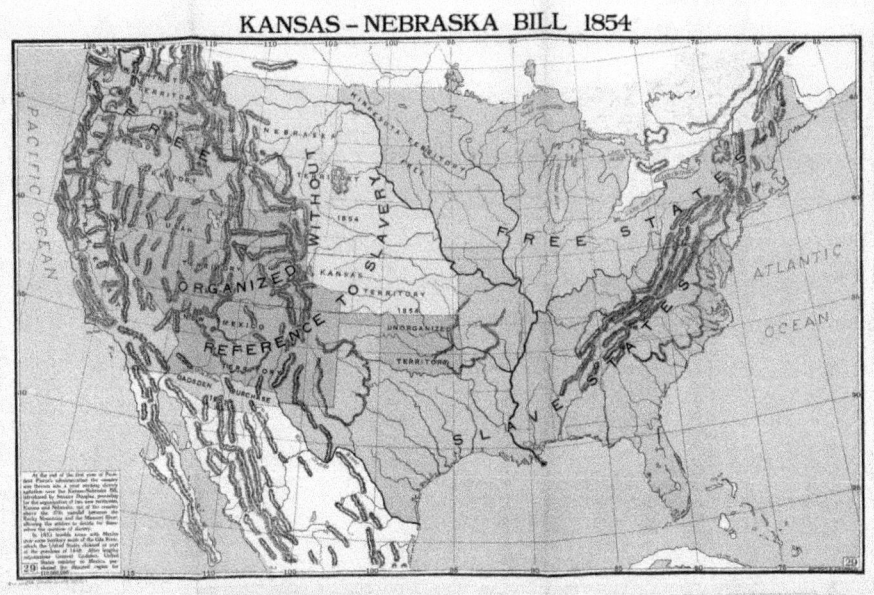

The Dred Scott Decision

One last hope remained. The pending case of Dred Scott vs Stanford (the name was Sanford, but the clerk mistyped the name) gave hope the issue of slavery might be resolved. They were grasping at straws. Dred and Harriet Scott were slaves who had lived on free soil and, as a result, claimed they were free. The Court ruled (Taney Court refers to the Supreme Court from 1836 to 1864 with Roger Taney as the Chief Justice) that the case should never have been brought. Slaves were property, not citizens, and thus had no right to use the court system. The framers of the Constitution, Taney argued, did not consider slaves citizens. The Missouri Compromise was unconstitutional as it violated the Fifth Amendment which involved the depriving of property. Also, Congress could not bar slavery in any territory. That could only be done with the formation of a state by that state. Far from solving the slavery problem, the Taney Court ignited a firestorm. It might be argued that expecting the Supreme Court to solve this issue that had defied solution for so long was unrealistic. At this point, the prestige of the Taney Court sank to such a low that President Lincoln, during the Civil War, could actually threaten to have the Court arrested, and it faded into the background.

Final Straws

As the nation teetered on the brink of a civil war, it needed only a few more items to push it into that war. One occurred when radical abolitionist John Brown and some followers attempted to seize the federal arsenal at Harper's Ferry in Virginia. His aim was to arm a general slave revolt to march through the South. The raid was quickly put down, and Brown would later hang for treason, but the incident convinced many Southerners that Northerners were out to turn their world upside down. The election of 1860 would be a sectional election. Abraham Lincoln would win the north, Breckenridge the south, and Bell and Douglas the border states. Receiving only 40% of the popular vote, Lincoln would garner an Electoral College victory. Although only advocating the limitation of the spread of slavery at this point, the election of Lincoln led to the secession of South Carolina, followed by the other states of the Lower South. A flurry of Congressional "deals" were put forward, but none were successful. When President Lincoln decided to resupply the U.S. Army garrisoned at Fort Sumter in Charleston Harbor, the Fort was fired on April 12, 1861, by the South Carolina Militia. There were no casualties on either side – it was a mild start to a very bloody affair. The straw, tinder, and logs I discussed had finally caught fire!

Chapter Two

MISSOURI

The Civil War is often viewed as an "out east" thing. Indeed, you can visit many a battle site in Virginia that was a part of the "on to Richmond" drama. However, as you have noticed in my summation, the name Missouri pops up often. In fact, Missouri would fight a "Civil War" in a Civil War to determine which direction the state would go.

A Divided State in a Divided Nation

As the thunder of cannon fire roared in Charleston Harbor, the border state of Missouri found itself with mixed emotions and divided loyalties. More than two-thirds of her white population were of southern stock, while in St. Louis, 65,000 free-thinking German immigrants formed a core of Union support.

To Missouri, slavery was not of primary economic importance in comparison with other slave states. There were 115,000 slaves owned by only 24,000 planters and farmers. Above all, Missourians wanted compromise and peace; If the war came, they desired neutrality.

The state of Missouri was of vital importance both to the Union and to the Confederacy. Her substantial manpower pool, her strategic geographical position, her resources, and wealth were needed sorely by both forces. Through important citizens whose feelings were strongly tied to one side or the other, Missouri felt strong tugs and pulls when war did erupt. And as the nation became divided, so did Missouri.

The Battles at Patterson and Fort Benton

A tall, dignified man, born in Kentucky and whose roots extended back to Virginia, set in the Governor's mansion. He was Claiborne Fox Jackson, 54, and able politician and a secret secessionist friend of the South. He maneuvered the state legislature into setting up a convention to decide Missouri's future relationship with the United states. However, the 99 delegates elected were mostly pro-union, believing that secession would ruin the state's economy. On March 4th, 1861, the convention, meeting in Saint Louis, decided that there was "no adequate cause to impel Missouri to dissolve her connection with the federal union."

This was a blow to the governor and pro-southerners but welcome news to men like St. Louisian Francis P Blair, Jr. Organizer of the new Republican Party in Saint Louis, he was staunchly loyal to Lincoln and the Union. A young and energetic 40, he was active as a lawyer, editor, and congressman. Through his brother, Montgomery, a member of Lincoln's cabinet, his influence was felt in Washington.

The alignment of opposing forces was most evident in Saint Louis, the state's largest city. A Union home Guard unit, "The Wide Awakes," composed mostly of Germans, held regular drills, as did a pro-south group called the "Minute Men". Neither had arms, and this situation focused the attention of all on the federal arsenal located below the city.

Francis Blair asked Washington to send Federal troops to reinforce the small arsenal staff. An 80-man detachment was dispatched by steamboat from Ft. Riley. Their leader was captain Nathaniel Lyon, a career Army officer who hated secessionists. This made him popular with Blair, who was cool to the doubtful allegiance of general William Selby Harney, commander of the Army's Department of the West.

Three days after Fort Sumter was fired upon, President Lincoln called for volunteers. His request of Governor Jackson to furnish four infantry regiments was defiantly refused. This prompted the worried captain Lyon to smuggle a large portion of the 60,000 muskets stored in the Saint Louis arsenal across the river into Union Illinois.

Franklin Blair quickly offered Lincoln his loyal "Wide Awakes" home guards since Jackson would not furnish the Union with any Missouri troops. Lincoln accepted, and Captain Lyon began mustering in men and arming them with weapons from the arsenal. The arsenal and its precious stores were safe and had been a convenient source of weapons for a local army. The union definitely had the edge.

Over in Jefferson City, Governor Jackson ordered six days of training for the militia and urged the legislature to pass money bills to pay for the defense of Missouri. At Saint Louis, the militia camp, named Jackson in honor of the Governor, was commanded by a West Pointer, Brigadier General D. M. Frost. It was located between Olive and Laclede streets east of Grand in Saint Louis. Blair and Lyon watched the encampment fearlessly since it housed a mere 700 or so militiamen but with irritation. But when Lyon learned that howitzers and large guns taken from a federal arsenal in Louisiana had been smuggled into camp Jackson, he took action against it. On May 10th, 1861, Captain Lyon road at the head of the home guard and surrounded the southern camp. He demanded surrender within a half hour and denounced it as a nest of secessionists. General frost displayed intelligence in offering no resistance.

The militiamen stacked arms and marched as prisoners of war to the arsenal under the guard of the "Wide Awakes." Crowds gathered to learn what had happened. Many in the mob sympathized with the militiamen, and soon pushing and shoving grew into rock throwing and then pistol shots. The union volunteers fired into the crowd, and some estimates placed the dead at 28. Missouri had shed some of the first blood to be let in the Civil War.

The capture of Camp Jackson sank the stock of the secessionist group in Saint Louis, ending its aggressive action there. The Confederate flag that flew from the roof of the Berthold mansion at Fifth and Pine streets headquarters of the "Minute Men" came down forever. Saint Louis itself was safe for the union.

The Battles at Patterson and Fort Benton

The secessionists at Jefferson City were thrown into turmoil by the news of camp Jackson's demise. The legislature passed appropriations to build and support a large state militia in what might be considered a record time of 15 minutes. Rumors ran rampant around the capitol. One, that Blair and several thousand troops were routed via the Pacific Railroad caused Jackson to have the railroad bridge over the Osage River burned.

General Harney, sent on a convenient mission so that captain Lyon could capture Camp Jackson, returned to Saint Louis. He made a pact with Governor Jackson that was very unpopular with Blair, Lyon, and loyal Unionists; the state would not arm further.

Francis P. Blair would not hold still for this concession. He used his Washington influence to have general Harney removed. His friend and fellow defender of the Union, Captain Lyon, was placed in charge, and was jumped in rank to Brigadier General. Governor Jackson thought that they should talk, and a meeting was called, though Lyon was cold to the idea. He had earned the sobriquet of "that Yankee abolitionist" among the more rabid Southerners and truly hated the Confederacy.

The meeting, attended by Lyon and Blair, Jackson, and his state militia head, Major General Sterling Price, took place at the Planters' House hotel in Saint Louis. Jackson proposed that Missouri maintained neutrality, under which he would disband the state militia and Lyon would refrain from enlisting volunteers and making troop movements in Missouri.

Lyon would not agree to this proposition which he saw as a surrender of Union control of the state. "Rather than concede to the state of Missouri the right to demand that my government shall not enlist troops within her limits, or bring troops into the state whenever it pleases, or move troops at its own will into, out of, or through the state; rather than concede to the state of Missouri for one single instant the right to dictate to my government in any manner however unimportant, I would see you, and you, and you, and every single man, woman, and child in the state, dead and buried." This zealous believer in the Union could feel as strongly as the Confederate believers in state's rights. He closed the meeting by telling

Jackson summarily, "This means war. In an hour, one of my officers will call for you and conduct you out of my lines."

On the return trip to Jefferson City, Jackson's party stopped to burn the railroad bridge over the Gasconade. Although the train arrived in the state capital in the middle of the night, Governor Jackson immediately drafted a proclamation calling Missourians to arms.

He asked for 50,000 volunteers for the state militia. It served to arouse Lyon to action back in Saint Louis. He was determined to break up any military coup that would carry the state into the Confederate camp.

The next day, he loaded a detachment composed of both regulars and volunteers aboard two steamboats headed up the Missouri River for Jefferson City. They had artillery aboard with which to lay siege to the capitol if necessary.

Since less than 150 men had answered his call to arms, Jackson realized the defense of the capitol was an impossibility. He grabbed the great seal of Missouri and scampered to Booneville, a strategic point located on high ground 50 miles upriver. General Sterling Price concentrated his state militia troops there and prepared for battle. Price was a huge man, a Virginian by birth who had settled down to farming in Charlton County, Missouri he, like Lyon, had fought with distinction in the Mexican war. Twice elected governor of Missouri, he had voted pro-Union as head of the state convention. When the breach came, he had to cast his lot with the South. He was a popular man, known as "Old Pap" to his troops.

General Lyon and his army occupied Jefferson City, then quickly moved by boat to Boonville to engage the state militia. His artillery opened fire on the Confederate positions at 8:00 a.m., then his well-trained troops advanced. The half a thousand militiamen broke and ran. The battle lasted half an hour with both sides inflicting light losses. The War in Missouri had begun in earnest, with the tides of fortune favoring the Union. This first skirmish was to be followed by nearly 1100 more, roughly one-sixth of the entire number to be fought in the Civil War. Each side would have its victories, each its losses.

The Battles at Patterson and Fort Benton

General Price's Lexington units were ordered south immediately by Jackson. He moved in that direction with the Boonville defenders, planning to rendezvous with General Ben McCulloch's Arkansas Confederates. General Lyon and another column dispatched from St. Louis followed and occupied Springfield. Union colonel Franz Sigel, the European train strategist, had been sent to Neosho to halt the flow of state militiamen slipping south to join the Confederates. Sigel moved north to intercept Jackson camped near Lamar, leaving about 90 men at Neosho. On July 5th, 1861, Sigel and Jackson met north of Carthage.

Though surrounded by State Guard cavalry on his flanks, the wiley Sigel disengaged his forces and ambushed the guard's infantry attack. He made a fighting retreat to Carthage. Here, he stubbornly defended, then slipped away south and east. His return to Neosho found those men engulfed by Confederates from the Arkansas border area. Thus, the State Guard, commanded by Governor Jackson had won a battle in the Missouri war.

Meanwhile, General Lyon at Springfield realized he was outnumbered and sent General Fremont an urgent plea for reinforcements. It went unanswered. Lyon then decided to pull a surprise attack on the combined forces of Generals Price and McCulloch rather than retreat. His force made a night March on August 9 to Wilson's Creek, where the Southern force was encamped. His plan was simple; Colonel Sigel's group would attack the rebels from the rear, then he would launch a frontal assault. Sigel was beaten off, retreating to Springfield without notifying Lyon. This capable officer engaged 13,000 enemy troops with his own remaining 3,700. Gaining a hill, he beat off numerous assaults by the Confederates. The battle, one of the bloodiest and most furious in the War considering the number of men engaged, lasted four hours. Near the end, General Lyon, already twice wounded, was killed. When the lull came, his successor, Major Samuel D. Sturgis, ordered a Union retreat. The Confederates, ready to do the same, were exuberant. Then Price hit upon a unique idea to protect his charging troops. Huge bales of hemp became rolling breastworks as his troops stormed the hill and moved on to the college.

Mulligan surrendered. Price took the entire Garrison prisoner and gained considerable supplies, horses, and arms. He also dug up the treasure buried previously by Governor Jackson. In winning this important battle, Price lost few men; 25 killed and 75 wounded; Mulligan had 39 killed and 120 wounded.

With the victory at Lexington, secessionist's hopes reached their peak in Missouri, never to be this high again. Governor Jackson, though deposed, called a part of his deposed legislature together at Neosho, and on October 28th, 1861, it passed an act delivering the state into the confederacy. The union scoffed, but the confederacy admitted Missouri as its twelfth member.

Severe criticism of General Fremont, head of the Department of the West, issued from Missouri and Washington. "The Pathfinder" Fremont, eager to save face, moved with 50,000 men to personally take on Price. Just when he had Price trapped in southwest Missouri, he was relieved of his command. Under instructions, the new commander, general David Hunter, pulled back to Saint Louis. Again, the state was Price's as he regained much of western Missouri with an army of 15,000 an initiated Guerilla warfare with unorthodox but effective tactics.

In southeast Missouri, a Union General, Ulysses S. Grant, soon to gain stature and reputation, commanded from his headquarters in Cairo. Fremont had put him in command of southeast Missouri. Grant made his headquarters in Cairo, Illinois. When Confederate General Leonidas Polk violated the neutrality of Kentucky and set up camp in Columbus in that state, Grant countered and occupied Paducah. Besides blockading the Mississippi from its fortifications at Columbus, Polk intended to take St. Louis via an invasion of southeastern Missouri. He sent General W. Jeff Thompson down the river from Columbus to operate around New Madrid, Missouri. Thompson started pushing and probing north in October 1861 with a force of 2,000 men. Grant sent troops from Pilot Knob and Cape Girardeau to shove him back to the marshes of New Madrid.

Next, Grant turned his sights on a Confederate camp of 2,500 men at Belmont, Missouri, across the river from Polk's headquarters at Columbus.

The Battles at Patterson and Fort Benton

Taking 3,000 troops aboard steamboats, he attacked the unsuspecting Confederates and put them to flight. But his green troops were careless and began looting the camp. The enemy regrouped, and the siege guns at Columbus bombarded them, too. The Union troops fought their way out, but Washington frowned on Grant's action.

As the North's military potential grew, the South's shrunk everywhere, including Missouri. General Henry Wager Halleck had replaced Fremont as head of the Department of the West. His efficiency began to shift the tides of war in favor of the Union. Price found himself forced to retreat to Springfield, where he remained during the winter of 1861-2. Halleck sent an army of 15,000 under General Samuel R. Curtis to destroy Price's army or chase it into Arkansas.

In spite of a severe winter, Curtis did a workmanlike job of routing price out of Springfield south into Arkansas. He then chased General Ben McCulloch out of his winter quarters at Cross Hollows, Arkansas, and occupied it himself. Curtis had stretched his line of communications thin. He also found the Confederates in Arkansas able to join forces. Price had 5000 Missourians, McCulloch about 15,000 Confederates, and General Albert Pike, a character who was poet, politician, and friend of Indians, had 5,000 redskins in war paint ready to fight for the South. Price and McCulloch could not get along, so General Earl Van Dorn, their boss and commander of Trans-Mississippi Department Number 2, took personal charge of this combined army.

Van Dorn eagerly set out to attack Curtis, who had decided to withdraw behind Sugar Creek, just South of the Missouri border, and meet the Confederate onslaught there. Curtis's four divisions, less 2 regiments of Colonel Franz Sigel's command, which were lost to Van Dorn in skirmishes on the withdrawal, dug in with Pea Ridge at their backs. They waited for tomorrow's fight.

The morning found the Confederates not at their front but in their rear. Van Dorn was executing a double envelopment from alongside and behind Pea Ridge. Curtis had his units do an about-face, and his front became his rear, his rear his new front. On this day, March 7th, 1862, the numerically superior but tired and hungry Confederate force pushed the Union troops back; Pike's Indians nearly frightened one Union division to death, and Price's Missourians pushed a stubborn Federal division south in the vicinity of a country inn named Elkhorn Tavern. But before day's end, McCulloch had been killed by an Illinois sharpshooter, Private Peter Pelican. His second in command, General McIntosh, was killed shortly afterward, and this wing became demoralized. Pike's Indians ran, tired of fighting white man's fashion.

Van Dorn's army started the next day with an ammunition shortage and one day hungrier and colder. His numerical superiority was also greatly reduced. Curtis, who had refused to follow his division commanders' council calling for a retreat, ordered Sigel to attack with his two divisions. The German's deadly artillery fire smashed the Confederate batteries and the infantry's morale. Van Dorn's force fell apart, retreating wildly in every direction but south. It took two weeks to reassemble the beaten army at Van Buren. Curtis had defeated a superior force with cool skill in a battle that was decisive in saving Missouri for the Union.

The Guerrilla Years: A War of Revenge

After the smoke of battle had cleared in Missouri in 1861 and the War proper had rolled on south, the state found itself the scene, not of peace, but of another type of war, unique in character. It was a private war, with Missourian usually pitted against fellow Missourian, neighbor against neighbor. This was war as practiced by guerrillas, civil war in the precise sense of the term; A bloody, vengeful, vicious community insurrection.

The number of participants engaged was small, and battles were raids, skirmishes, and ambushes. Yet it took a terrific toll on life, and unlike the war fought by the armies, there was no respite, no rules, and no military objectives other than those close at hand. It was a war of terror, surprise, sabotage, and arson. It was kindled in the spirit of bitterness and revenge in the breasts of Southern sympathizers, particularly those subjected to Union military law or occupation.

The Battles at Patterson and Fort Benton

The outburst of this strange type of war in Missouri was caused by factors social, political and military in nature. In western Missouri and in the counties along the Missouri River, slavery was strong. The entire state was predominantly southern in background and outlook. The state was somewhat of a contradiction; Its people voted almost unanimously to maintain the Union, while its governor, a secret secessionist, refused to furnish troops for the Union cause. The North viewed Missouri as a slave state and a stronghold of secessionists. This was fostered earlier by the illegal and violent attempts in 1854 of some of its citizens to make Kansas a slave area.

Thus, when the War began, Northern forces pushed the secessionist state militia beyond its borders; the Union occupation troops, mostly from neighboring Illinois, Kansas, and Iowa, treated and thought of Missourians all as secessionist and disloyal to the federal government. This caused excesses of martial law and abuses in military government during the occupation to occur, creating resentment in the natives. The Union occupation bisected the state at its center, leaving many citizens disloyal to it behind its lines. It was these men who instituted the art of guerrilla, or partisan, warfare.

Organized into small, mobile cavalry groups, they permitted no peace in central, southern, or western Missouri for the duration of the War. They plagued the Union forces stationed in the state and, in 1862, caused complete mobilization of the state's manpower. Noting the success of these local units, although not in sympathy with their disregard for the laws of civilized warfare, the Trans-Mississippi Department of the Confederacy and General Price sent guerrilla parties from Arkansas into Missouri to recruit and destroy Union transport facilities and communications. Joining forces with the Missouri groups, they created havoc and horror in unabating fury for four solid years.

Although the large-scale actions of the organized forces far overshadowed the activities of the guerrillas, their role was an important one. The Missouri Confederate guerrillas were highly effective and efficiently utilized in a peculiar type of combat. They proved that by unorthodox tactics, a smaller force could keep an overwhelmingly larger force off balance, mobilized and tied up, unavailable for utilization in military campaigns elsewhere. Their activity against the civil population created such fear and disorganization that society collapsed in some areas. Evidence of their effectiveness was the notorious order number 11 issued by Union authorities, which drove the residents of four border counties from their homes. This was an attempt to stamp out guerrilla resistance in the area. These irregular bands actually caused martial law to be imposed, with this suspension of the legal and constitutional rights of the state's citizens. In short, they fought a cruel, total war.

A guerrilla is a much different type of fighting man than a disciplined soldier who is a member of a professional army. The Missouri Confederate guerrilla either fought for immediate, emotional reasons, such as revenge for the treatment accorded his family, hatred of those who burned his home, or out of sheer criminality, recklessness, or neuroticism. He had no rules of conduct; He gave no quarter and expected none. In order to succeed in his missions, in fact, to survive, he needed to be a woodsman, horseman, and pistol shot of first-rate proportions. Many were underage for military service but with a need for excitement; Farm boys of a high type were eager to play a deadly game of chance over the rolling, wooded sectors of western and central Missouri.

Guerrilla leaders were unusual men; Some of great abilities and capacities, and others were misfits in any other type of life. Able cavalry leaders all, their daring and viciousness have earned a niche for them in American history, folklore and romantic fiction and movies. They are better remembered than the leaders of both North and South who took part in the Civil War in Missouri. To Missourians contemporary with them, they were highly controversial figures. Citizens with Southern leanings made them into heroes of gigantic stature, equal in mark of heroes of old. To the union sympathizers, they were contemptuous characters, unprincipled outlaws, and "bushwhackers." Indeed, they were colorful, some even bizarre.

The Battles at Patterson and Fort Benton

The most notorious Missouri guerrilla captain was William C. Quantrill. He was, oddly enough, a northerner, a native of Ohio. Prior to the war he had taught school in Kansas. To secure himself from attack as a turncoat and informer, he manufactured a background for himself that he had come to Kansas from Maryland en route to Pikes Peak but had been attacked by James Montgomery's pro-union Kansas Raiders and had been wounded. After recovering from his wounds, he joined Montgomery's Jayhawkers under an assumed name to gain revenge. Most of his men believed it.

His was the largest and perhaps the most active band. It was composed mostly of farm boys from western Missouri who ached for revenge against the harsh treatment accorded their relatives by the Kansas Union troops who occupied the area. In the summer of 1862, his ranks were swelled by ex-members of General Price's State Guard, who, after being paroled, found themselves suspect by Union authorities. They and their families received brutal treatment from the Union troops. Upon joining Quantrill's organization, a guerrilla was asked one question only; "Will you follow orders, be true to your comrades, and kill those who serve and support the union?"

Quantrill was a competent leader but was amoral, vicious, and highly ambitious. Into his command came lieutenants like George Todd, a bridge Mason from Jackson County. He was a possible psychopath and later led a group of his own. Another Quantrill follower was "Bloody Bill" Anderson, a wild and emotional raider. His terrible deeds were to exceed, by far, those of his leader. Whereas Quantrill was a calm, nerveless and calculating man when approaching a skirmish, Anderson attacked crying and screaming. He was an extremely handsome man, tall, sinewy and lithe. His dark hair was curly, and it fell to his shoulders. Anderson had high cheekbones and large, piercing blue eyes that literally blazed. He was an elegant dresser and made a dashing figure on horseback. His emotional makeup was definitely unstable; He was volatile and violent.

Anderson was a native Missourian, reared and educated at Huntsville, the county seat of Randolph County. As a young boy, he moved with his father, his brother Jim and three sisters to Kansas, where the family was at once caught up in the border war. Bill's father was killed, and some forgotten border foray and he and his brother joined Quantrill. When one of his young sisters was killed and another injured in the collapse of a Union military prison in Kansas City in 1863, he became motivated by an insane desire for revenge. He killed every Union soldier and every civilian who supported the Union cause that fell into his hands.

Anderson had as his Lieutenant, personal bodyguard, and executioner a boy as neurotic and unbalanced as himself. This underlings name was Archie Clement who had been reared at Kingsville in Johnson County, Missouri. The eighteen-year-old Clement was a gunman in the true border sense of the word. Small, blonde, grey-eyed, he wore a perpetual twisted smile. He was completely lacking in feeling, scalping and mutilating his victims cruelly with a savage lust for violence. The only authority he knew or obeyed was "Wild Bill" Anderson, to whom he was devoted. Anderson needed but give the word, and Clement was killed with a pistol or knife.

Another Quantrill recruit whose motive was revenge was 14-year-old Riley Crawford. His father, Jeptha Crawford, had been taken from his home near Blue Springs in Jackson County and shot by Kansas Jayhawkers. Mrs. Crawford then decided to use her boy as a tool of vengeance and delivered him to Quantrill's camp. She asked the guerrilla leader to make a soldier of the lad. Until he was shot dead at age 16 in Cooper County in 1864, little Riley killed every Union soldier who fell into his hands.

Some of Quantrill's men achieved notoriety after the War, too. One of the State Militia's parolees who signed up with Quantrill was Frank James of Clay County. 19 years old and rail thin, Frank soon sent for his 17-year-old brother, Jesse Woodson James. Both were daring horsemen, wild and hard riding on guerrilla expeditions. Later, they were to write their names in blood in our nation's folklore and go down in history as legends. Another pair of brothers who would continue their infamous acts in the peace that followed the War were the Youngers, Cole and Jim. Coleman Younger was eighteen when he entered Quantrill's camp in 1862. The year before, his father's fortune had been stolen at Harrisonville and carried off by Kansas Jayhawkers. Soon after he had joined Quantrill, Younger's father had been brutally robbed and murdered by a Union officer. Before the end of the War, the Younger home was burned, and Cole's mother and the rest of the family turned out into the winter by Union troops. In 1864, Cole's 16-year-old brother Jim joined him to seek their revenge riding with Quantrill. After the war, they joined the James brothers as outlaws.

Quantrill's band helped capture the town of Independence, Missouri, and the Union force garrisoned there on August 11, 1862. For this, Quantrill received a captaincy in the Confederate Army. On August 23, 1863, he led his guerrilla troops into Lawrence, Kansas, where they literally sacked the town, burning it to the ground. Their viciousness and thirst for revenge was proved by the ruthless, needless murder of some 150 men and boys in the town. In October of the same year, the band defeated a small Union cavalry unit at Baxter Springs, Kansas, and put 17 captured non-combatants to death. Other guerrilla massacres occurred at Concordia and Centralia in Missouri.

As the Civil War drew to a close in Missouri, the Union military command slowly but surely exterminated most of the guerrilla leaders and their followers. George Todd was shot from his horse at the Battle of Independence in 1864 while leading a scouting party of General Joe Shelby's Brigade. "Bloody Bill" Anderson was killed in a skirmish in Ray County a few days later, and his body was triumphantly carried to Richmond and photographed. Archie Clement survived the war but was assassinated at Lexington, Missouri, in 1866. Quantrill, who had entered Kentucky in 1865, looting and robbing there until May of that year, was surprised by a small Union force and was wounded. He died in a Louisville hospital.

The Confederate guerrilla forces, ranging the border from Missouri to Texas, were engaged in an important, if unique, facet of the Civil War. Theirs was a role undignified by uniforms, military codes, and master strategy. Instead, they fought a total War, very personal in nature, cold-blooded, using unorthodox hit and run tactics. In waging this sort of war, they not only avenged many personal wrongs done to them and theirs but made a real contribution to the Confederate war effort in the West. For three years, they kept Missouri embroiled in a private war, a partisan war fostered by an unfortunate political and military situation that might have been better handled by the Union authorities. The fierceness and fury of this war appalled all Missourians and touched most of them personally.

The Dying, Desperate Gamble That Lost

Two and one-half years after the Civil War had swirled south and east of Missouri, it returned to sweep across the state with a final fling by the Confederacy. A veritable ghost was at its head, Major General Sterling "Old Pap" Price. He dreamed of an invasion of Missouri that would capture St. Louis and then drive the Union forces from its soil. He believed Missouri housed many persons with Southern sympathies who would act if an organized effort could be mounted. He counted highly on the aid of guerrilla bands, then marauding throughout the state. He argued that his new front would draw Union forces from other beleaguered areas, taking the pressure off of them. Underneath, he also smarted from his earlier defeats, and his thinking was tinged with vengeance.

Gaining approval for his project, he assembled three columns totaling 12,000 mounted infantry and cavalry troops supported by 14 pieces of artillery. These were headed by General John S Marmaduke, General James F. Fagan, and famed cavalryman General Jo Shelby. The troops were mostly from Missouri and Arkansas. Price took with him Thomas C. Reynolds, former Lieutenant Governor under Jackson and head of Missouri's Confederate government in exile. This tattered, tired force crossed into Missouri on September 19, 1864, converging on Fredericktown. From this small town in southeast Missouri, Price's army now rode West, headed for Jefferson City. Its first objective was Fort Davidson, situated in a gap in the Ozarks at Pilot Knob, 85 miles South of Saint Louis. General Thomas S. Ewing had 1000 men, half of them green recruits, with which to defend the fortress. He had built a moat, sturdy earthworks, and rifle pits about it.

The Battles at Patterson and Fort Benton

General Fagan's column attacked in a frontal assault, to be met by withering fire from these inexperienced but determined Union troops. Fagan's column was stopped cold, and a second try saw his force completely humbled. Next, General Marmaduke's division was hurled against the fort from another position, but it received the same kind of treatment. Then Price ordered a pincer move by both units, but neither jaw could snap shut, and this attack petered out at the edge of the deep moat. Changing his strategy, Price hauled his artillery pieces to the tops of the two mountains that overlooked the fort and prepared to reduce it by bombardment. Ewing's force quietly slipped out that night, joining the strong Union force at the Rolla base of operations.

Meanwhile, General W. S. Rosecrans, head of the Department of the West, was busily throwing up defenses to protect St. Louis. He was compelled to concentrate forces at several key points around the state since he could not be sure where Price would strike. St. Louis itself was practically defenseless since the War had swept south. Quickly, Rosecrans formed a division of 4,500 home guards. He was then reinforced by 6,000 troops under the command of General A. J. Smith on their way to join Sherman in Georgia.

Price's mounted army moved on toward Jefferson City, shaken by its losses at Fort Davidson. Few recruits had joined him to take the places of those lost in that costly attack. He moved West, leaving in his wake a 20-mile-wide swath of destruction. Waiting at Jefferson City were 7,200 Federals, entrenched and ready for battle. Price held up and engaged this force in a one-day skirmish on October 7th with a half-hearted effort and then decided to bypass the capital city. This was a great disappointment to Thomas C. Reynolds, who had waited longingly for his triumphant return to Jefferson City and installation as governor.

Union forces were now in pursuit of Price's dwindling army of 9,000 men. Daily skirmishes were fought. Guerrilla bands under Quantrill and "Bloody Bill" Anderson joined him at Boonville. More battles at Glasgow and Lexington were fought and Price then neared Kansas City. Still full of fight, his troops pushed a Federal line of Kansas militia back from the Little Blue River to more permanent and formidable positions behind the Big Blue River on October 21, 1864. The next day, Jo Shelby forced his way across the Big Blue, and the Union line withdrew to prepare defenses in front of Westport in Kansas City.

The following day, Sunday, October 23, was a bloody day as 29,000 troops engaged in the Battle of Westport. General Curtis, hero of Pea Ridge, led 15,000 Union soldiers; General Alfred Pleasanton another 5,000. Pleasanton attacked from the rear, destroying General Marmaduke's holding force. General Shelby had driven the Federals off his hill time and again but was finally overwhelmed. Price's army was in retreat, but he rallied them for a second stand. It crumbled, and retreat in wild confusion ensued. Price was pushed back along the Missouri-Kansas border all the way south of the Arkansas River. The ill-starred campaign was ended, and with it, the Civil War in Missouri.

Chapter Three

FORT BENTON FELL TWICE TO THE SOUTH IN THE UN-CIVIL WAR – BY DAVID HAGLER.

The civil unrest of 1861 found most southern Missouri citizens living in rural areas, far from neighbors and far from the political upheaval of government. News trickled ever so slowly. Washington was at the other end of the earth.

Tending to one's family, tending to one's farm or trade were the important things- then tending to one's own business. They thought they were far away, removed from all the shouting and shoving going on elsewhere.

Many of the people that populated this area had migrated from states in southern regions, so they sympathized with whatever affected their heritage, whether they understood the political complexities or not. To the Federal Military Command posted at St. Louis, this region was known as "sesish" land for its' presumed secessionist views and was considered hostile for the most part.

A Federal post was proposed in this quarter, a listening post, that might signal a first strike made by a Confederate force moving from the south. In 1861, a post was established at Patterson, a few miles from present-day Sam A. Baker State Park. Patterson was connected by road to Ironton, 30 miles to the north. Near Ironton was a large Federal military post at the end of a rail line originating from St. Louis, where supplies and communications flowed.

Patterson was a thriving community the local residents referred to as "The Cross Roads,"; describing the intersecting roads in the center of town.

The one running north to Ironton also went south toward Greenville and the other, east to Cape Girardeau and west toward Springfield. The town was militarily valuable. So, William Patterson's fields were cleared. A hill approximately one-quarter mile south of Patterson gave a commanding view of the valley and these crossroads.

A small earthen fort was constructed, possibly built in 1862, after a base was established in 1861. As a common practice forts were named after a prominent figure. Fort Benton was named after William Plummer Benton, a Colonel of the 8th Indiana Infantry.

The Fort

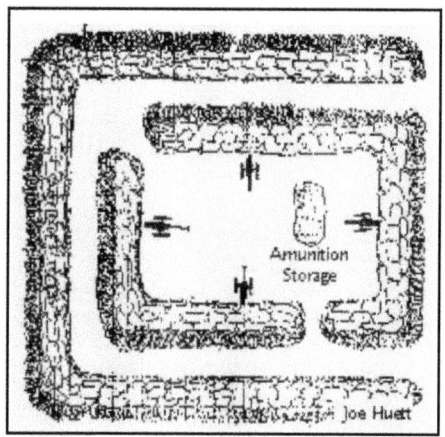

Fort Benton is an earthen fortification of square design. Each side faces a direction of the compass, each being approximately 150 feet in length, while the dimensions of the interior are approximately 100 feet per inside section. Trenches were dug around the exterior of the fort, and dirt was thrown on the top, creating earthen walls. There appears to have been a powder magazine in the center to facilitate using any artillery pieces that might be placed in the fort.

History has described the interior of Fort Davidson at Pilot Knob. Fort Benton, being a sister fort, shared some characteristics. Fort Davidson's interior walls were built of five-foot-tall, two-inch thick lumber planks. The outer parts of the walls were made of packed earth and were five feet high and five feet thick. Earth-filled gunny sacks arranged like brickwork formed the top of the wall, also known as the parapet. Inside the fort, along each of the walls, a series of steps led to the firing platform where riflemen could kneel and fire over the walls.

I believe Fort Benton had many of these characteristics, especially the wood interior. It has been noted that when Confederate General Joseph Shelby captured Fort Benton on September 22, 1864, his troops dismantled the fort and burned it, suggesting the fort could have been wood-lined.

Base of Operations

During 1862, the Union army controlled Wayne County with garrisons at Patterson and Greenville. Mostly, it was a period of bushwhacking, skirmishing, and scouting actions. During the spring and summer of 1862 there was much activity enrolling local men in Union army units. Local history notes that many men of Southern sympathies were conscripted into Union service, thereby protecting their families from Union reprisals. Those who chose not to cooperate felt the harsh realities of war. Many husbands, sons, and brothers just disappeared. Because so many men had left the area or were in uniform, the old men, women, and children faced a harsh life, surviving only to be plundered by both sides.

Federal Colonel Edwin Smart commanded the 3rd Cavalry Missouri State Militia stationed at Patterson. The Federal units at Fort Benton were a thorn in the side of Confederate movements in this area, and its large stores of provisions were a tempting plum. Many Confederate soldiers needed rifles and tentage and Fort Benton could help supply it if its 600 soldiers could be taken.

The Battles at Patterson and Fort Benton

In the spring of 1863 this was in the mind of Confederate Brigadier General John S. Marmaduke when planning his second raid into Missouri. General Marmaduke, with a command consisting of 5,000 men, eight old pieces of field artillery, and two light mountain pieces, planned to encircle Patterson and capture the town, the fort, and all supplies. Of the 5,000 men, only 3,800 were armed. Most carried shotguns or a squirrel rifle. Very few soldiers were armed with a military rifle.

Word reached St. Louis that a raid was imminent. The army had installed a telegraph line from Patterson to Pilot Knob and from there to the high command at St. Louis. Picket posts and Cavalry scouting units were ordered on Patterson's perimeter.

On April 20, 1862, Confederates captured many of these outlying Union posts, and a small Union force stumbled onto the Confederates, and a noisy fight ensued. Surprise was lost.

Colonel Smart ordered all supplies loaded on wagons. What they could not haul, they set afire. They retreated north toward Fort Davidson, with Confederate troops in hot pursuit.

Confederate Fort

Patterson and Fort Benton had now changed hands. A detachment of Federal Cavalry dashed into Patterson, seeking safety from pursuing Confederate cavalry, but was ingloriously captured. The Confederates put the fires out and saved what stores they could.

Colonel Smart's command, being pursued, set up a defense line at Stoney Battery on Big Creek, a boulder-strewn area 7 miles north of Patterson. They defended the creek crossing, giving time to the supply wagons escaping north with the valuable cargo. A brisk delaying action resulted in 23 Federals killed, 44 wounded, and 53 taken prisoners. Confederate casualties were light at three wounded. But Federal supplies reached Fort Davidson safely.

150th Anniversary of the Stoney Battery clash – 4/20/2013

The Confederates held the fort for awhile but eventually retired back to Arkansas.

The Battles at Patterson and Fort Benton

In August of 1863, Lieutenant Colonel William J. Preston's 4th Missouri Cavalry, Confederate, received information that there were 30 wagons of Union Commissaries near Greenville in Wayne County. They did not locate their primary objective but redirected their movements toward Fort Benton.

At Patterson, a settler's wagon loaded with goods was captured, along with a Federal Captain, three privates, the Deputy Provost Marshal, including an Enrolling Officer. Preston and his men stayed at Patterson for four or five hours, then returned south by way of Van Buren after learning that north of Patterson, two squadrons of Federal Cavalry and three companies of troops had occupied Stoney Battery and the bridge on Big Creek.

October 1863, Captain W.T. Leeper and Company L of the 3rd Federal Cavalry made a thrust from Fort Benton as far south as Smithville, AR.

Military activities continued from Fort Benton throughout the winter of 1863 and into the summer of 1864. Meanwhile, Confederate spies and bushwhackers kept the area stirred up.

In the fall of 1864, Confederate General Sterling Price planned a diversionary raid into Missouri. A raid into Missouri might hold Federal units there and deny reinforcements to be sent east from Missouri, taking pressure off of General Lee. If St. Louis could be taken with all its arms and provisions, the state might be saved for the Confederacy.

On September 19, 1864, General Prices' expedition crossed the line into Missouri. It consisted of 12,000 men and was divided into three columns. General Fagan's center column had 4,000 armed fighting men, including General Price and his staff.

General Marmaduke, with 3,000 men, was assigned the difficult route through the east Missouri swamps. The western column consisted of 5,000 men commanded by General Joseph Shelby. General Shelby sent advance scouts ahead to Doniphan and surrounding areas.

Doniphan, on Current River, remained a rebel outpost through most of the war. Federal troops took this personally when several suffered food poisoning after eating there. A group of 40 or 50 paroled Federal prisoners stopped at Doniphan one evening seeking shelter. The citizens refused and a hotel owner set his dogs on them. Almost nude, barefooted, and starving, they struggled on to Ironton. Angry, their leader declared if he ever got back to Doniphan, he'd burn the place.

The same day Price entered Missouri, a detachment of Federal cavalry from Fort Benton reached Doniphan at 5 a.m. to scout the rumored Confederate advance. After capturing two rebel pickets, the federals charged into Doniphan, firing their weapons.

Forty Confederate men fled, crossing the bridge over the Current River. They destroyed the bridge and waited for the Feds on the opposite side. The Federals found a crossing a mile above Doniphan and pursued the 40 Rebs back to Arkansas.

The men of the Missouri State Militia on this raid carried a special order. Doniphan was burned to the ground, including the hotel. Feeling the score was now settled, the Missouri State Militia set up camp a few miles north of the burned town.

That evening, Shelby sent a party of 150 cavalry to investigate the incident. Reaching Doniphan, they found the town in flames, with its inhabitants distraught and homeless. On hearing this news, General Shelby's anger knew no bounds.

In the morning, as the Federal Missouri State Militia mounted their horses, preparing to head back to Patterson, they were fired upon. During the night, they had been surrounded on all sides by Confederates. Springing up on their horses, they tried to break out, firing their weapons as they galloped for their freedom. But the Federals were pushed back; the attack was so intense. They became confused and divided; it was now each man for himself.

There are two different accounts: one has 100 men with losses of 47 killed and 43 captured; another, 86 men with 12 killed, several pursued, captured, and shot on the spot. After this encounter, Gen. Shelby pushed on toward Fort Benton.

The Battles at Patterson and Fort Benton

Captain Robert McElroy of Company D, 3rd Missouri State Militia, was in command at Fort Benton on September 22, 1864. On that morning, Shelby's forces surrounded Patterson. Its garrison had heard of Shelby's advance and tried to retreat hastily, but not before many Federals were killed or captured. Some supplies were taken. The telegraph station was taken over so swiftly that the operator never completed a message of warning to Pilot Knob.

All the military facilities were destroyed, along with Fort Benton. Twenty-eight Federal soldiers were killed and several wounded. After the attack General Shelby and his forces continued to their next directive, Fredericktown.

General Price and his column arrived at Patterson that same day and buried the dead Federal soldiers in an unmarked grave in the northwest corner of the Fort Benton Military Parade Grounds, now the Patterson Cemetery. The next day both Confederate armies moved toward Pilot Knob and Fort Davidson. There, perhaps buoyed by their success at Fort Benton, the Confederates attacked prematurely and endured the bloodiest battle west of the Mississippi.

Later, after Prices raid ended, Patterson remained an important Federal link in this area. General Ewing, a Union hero at Fort Davidson, stated that Patterson would be re-occupied and that efforts to locate Rebels and Rebel families would continue until the end of the war. Fort Benton stands silent today, not as a tribute to either side. It stands as a monument to the courage of the Ozark region.

Here is a picture of the Fort Benton earthworks.

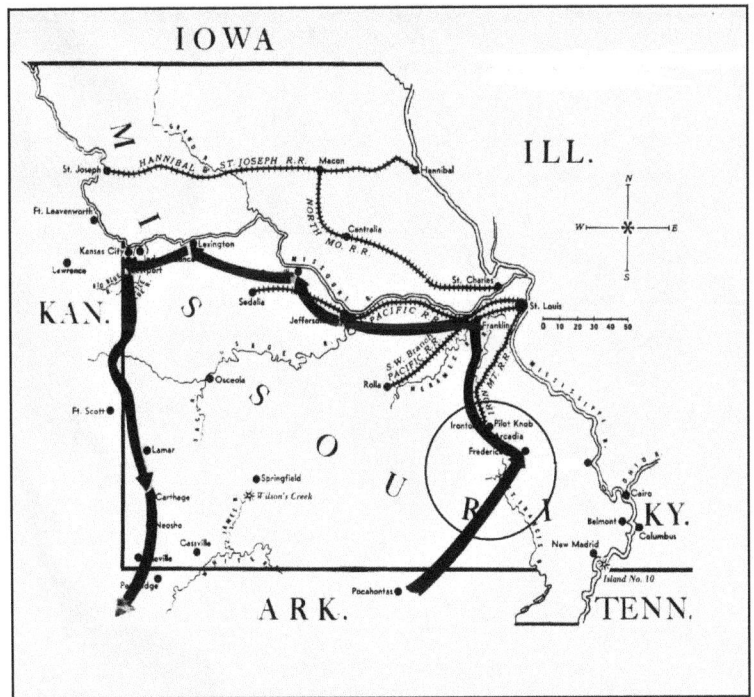

Route of the ill-starred invasion of Missouri by General Price

The Battles at Patterson and Fort Benton

A CHRONOLOGICAL LIST OF THE MAJOR CIVIL WAR BATTLES
and Events in Missouri — 1861-1865

Name of Battle	Date
Liberty, Seizure of United States Arsenal	April 20, 1861
Camp Jackson, St. Louis	May 10, 1861
St. Louis, Street Riot	May 11, 1861
Boonville	June 17, 1861
Independence	June 17, 1861
Farmington	July 4, 1861
Carthage	July 5, 1861
Neosho, Capture of Union Troops	July 5, 1861
Athens	August 5, 1861
Potosi	August 10, 1861
Springfield	August 10, 1861
Wilson's Creek, Springfield or Oak Hills	August 10, 1861
Birds Point, or Charleston	August 19, 1861
Lexington	August 29, 1861
Lexington, Surrender of by Union Forces	September 20, 1861
Osceola, Destruction of	September 22, 1861
Charleston, Expedition from, to Birds Point	October 2, 1861
Belmont	November 7, 1861
Warsaw, Destruction of United States Stores	November 21, 1861
Charleston	December 12, 1861
Mount Zion Church (Boone County)	December 28, 1861
New Madrid, Siege	February 28, 1862
New Madrid, Capture of	March 3, 1862
Clinton	March 30, 1862
Doniphan	April 1, 1862
Jackson	April 9, 1862
Bloomfield	May 10, 1862
Florida, Salt River	May 31, 1862
Lotspeich Farm, near Wadesburg	July 9, 1862
Moore's Mill, near Fulton	July 24, 1862
Kirksville	August 6, 1862
Newtonia	August 8, 1862
Independence, Surrender of Union Forces	August 11, 1862
Lone Jack	August 14, 1862
Lamar	August 24, 1862
Ozark, Captured by Confederate Troops	January 7, 1863
Springfield, at and near	January 8, 1863
Hartville	January 11, 1863
Bloomfield	January 27, 1863
Bloomfield near, and Capture of by Union Forces	March 1, 1863
Fredericktown	April 22, 1863
Cape Girardeau	April 26, 1863
Jackson	April 26, 1863
Sibley, Destruction of	June 23, 1863
Marshall	July 28, 1863
Boonville	October 11, 1863
Lamar, Destruction of by Confederate Forces	May 28, 1864
Laclede, Descent on	June 17, 1864
Fayette, near and at	July 1, 1864
Camden Point	July 13, 1864
Versailles	July 13, 1864
Huntsville, Attack on	July 15, 1864
Arrow Rock, Attack on	July 20, 1864
Plattsburg, Attack on	July 21, 1864
Shelbina, Attack on	July 26, 1864
Rocheport, near	August 20, 1864
Steelville	August 31, 1864
Tipton, Attack on	September 1, 1864
Centralia, at or near	September 7, 1864
Doniphan	September 19, 1864
Keytesville, Surrender of	September 20, 1864
Patterson	September 22, 1864
Farmington	September 24, 1864
Fayette, Attack on	September 24, 1864
Jackson	September 24, 1864
Arcadia Valley	September 26, 1864
Ironton	September 26, 1864
Shut-In Gap	September 26, 1864
Arcadia	September 27, 1864
Centralia	September 27, 1864
Fort Davidson, Pilot Knob, Attack on	September 27, 1864
Mineral Point	September 27, 1864
Franklin (Pacific)	October 1, 1864
Union	October 1, 1864
Washington Occupied by C.S.A.	October 2, 1864
Osage River	October 5, 1864
Jefferson City, at and near	October 7, 1864
Moreau Creek	October 7, 1864
Boonville, at and near	October 9, 1864
California	October 9, 1864
Danville, Attack on	October 14, 1864
Glasgow	October 15, 1864
Paris, Surrender of	October 15, 1864
Sedalia	October 15, 1864
Carrollton, Surrender of by Union Forces	October 17, 1864
Big Blue, or Byrams Ford	October 22, 1864
Big Blue	October 23, 1864
Westport	October 23, 1864
Charlot, or Marmiton	October 25, 1864
Clinton, Attack on	October 25, 1864
Newtonia	October 28, 1864

Chapter Four
HISTORY OF THE BATTLE OF PILOT KNOB

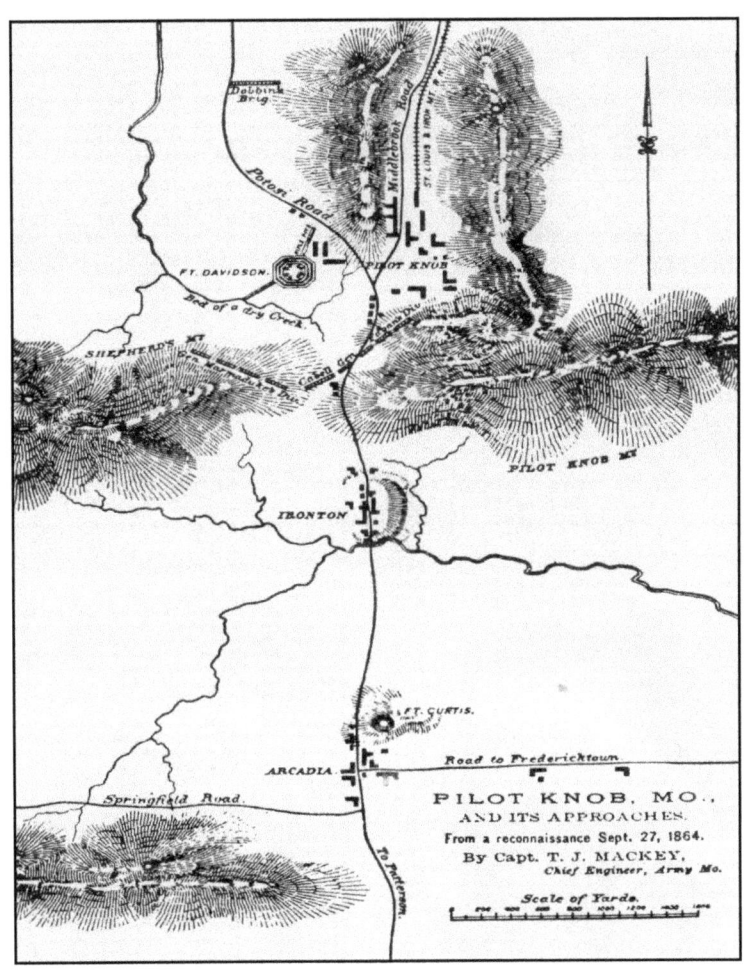

The Battles at Patterson and Fort Benton

The state of Missouri withstood more than 1,000 clashes during the Great Civil War. Only two other states, Virginia and Tennessee, had to endure more.

On the afternoon of September 27, 1864, the peaceful, picturesque Arcadia Valley of southeast Missouri was witness to one of the bloodiest battles of the entire war. In the brief span of 20 minutes, more than 1,000 officers and men lay wounded at the foot of Pilot Knob Mountain.

The Desperate Summer

By the summer of 1864, the fiery campaign of the Confederate States of America was on the verge of being snuffed out. East of the Mississippi River, General Ulysses S. Grant's Federal Army had the Southern army of General Robert E. Lee under siege and pinned down in Virginia.

Another Union Force under General William Tecumseh Sherman had stepped into the Confederacy's tough western army and was marching through Georgia to threaten Atlanta.

West of the Mississippi, in the Confederacy's Trans Mississippi Department, Commanding General Edmund Kirby Smith was behaving like the ruler of an empire separate and independent from the rest of the Confederate States.

Smith held the prize infantry on a tight rein, using them only to secure the borders of his sprawling domain. The lack of a Confederate military threat in the west allowed the Union to lightly garrison its western flank and to concentrate its strength in the east.

He was asked to send his best infantry east for the relief of Georgia and Alabama. Fearful of losing his infantry, Smith notified President Jefferson Davis that he was making plans for a major western campaign, which would be stymied by the loss of his infantry.

By August, Davis deferred to Smith, and not a single rifleman was sent across the Mississippi to the aid of Atlanta. The threatened loss of his infantry forced Smith to hastily arrange a raid into Arkansas and Missouri.

Smith chose Major General Sterling Price, a former governor of Missouri, to lead the western campaign. From the outset, however, it was clear that Smith did not intend to gamble many of his organized troops on an ambitious venture.

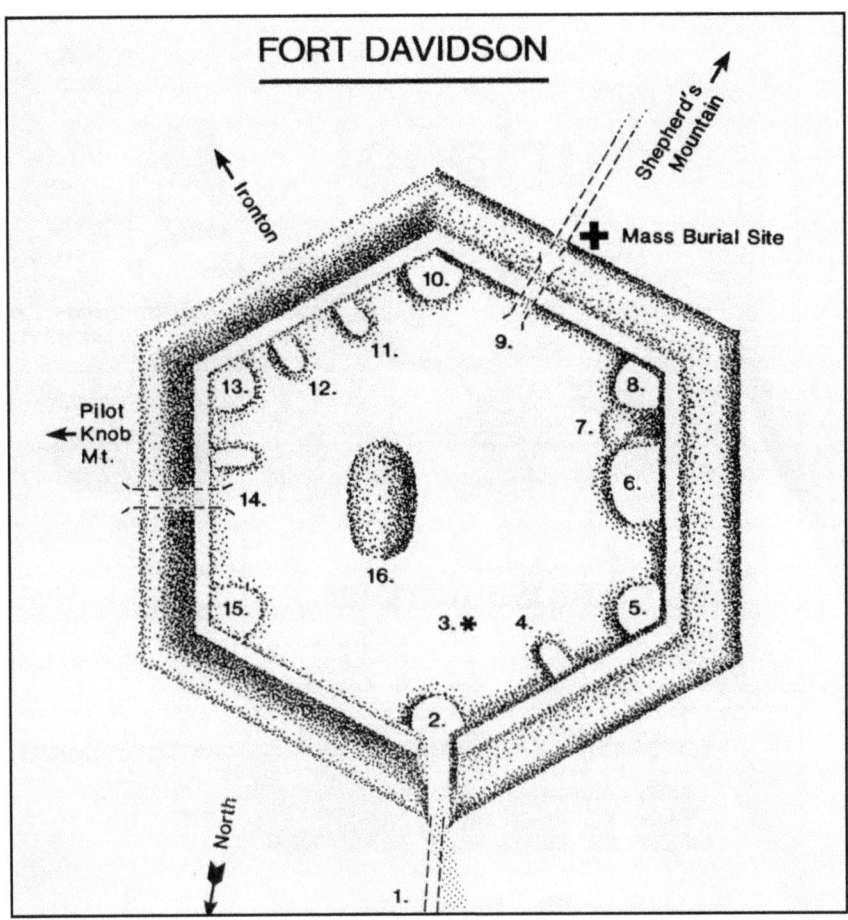

1. NORTH RIFLE PIT (no longer existing): From here, a trench ran 190 yards north to the town of Pilot Knob. Federal Troops defended it until just before the third Confederate assault. In the early hours of Sept. 28, the Federal infantry used this rifle pit to evacuate the fort unseen.

2. MODERN ENTRANCE: This entrance did not exist in 1864. From this earthen platform, a 24-pounder howitzer fired over the wall.
3. GENERAL EWINGS HEADQUARTERS: Brigadier General Thomas Ewing directed the battle from a tent erected at this point.
4. ARTILLARY PLATFORM: A 3-inch Ordnance Rifle of Battery H, 2nd Missouri Light Artillery, fired from this position. Platform to mount four of Battery H's six guns were erected during the night before the assault on the fort.
5. ARTILLARY PLATFORM: A 32-pounder siege gun fired from this position.
6. ARTILLARY PLATFORM: A 24-pounder howitzer fired from this position. The platform was enlarged to make room for a 3-inch Ordnance Rifle of Battery H, 2nd Missouri Light Artillery.
7. BORROW-PIT: This depression probably represents a place where the earth was excavated to build platforms for the guns of Battery H.
8. ARTILLARY PLATFORM: A 24-pounder howitzer fired from this position. The creek, just beyond the trees across the field, was the closest point reached by many of Marmaduke's Confederates. From there, they fired at the gunners who were exposed from the waist up.
9. SALLY PORT AND SOUTH RIFLE PIT: (No longer exists) Here, a tunnel, or "sally port," gave success to the ditch. During the evacuation of the fort, the Federal infantry marched out by the sally port, through the ditch, and down the north rifle pit. From a point on the other side of the ditch, the south rifle pit ran 150 yards toward the creek bed. It was abandoned before the first assault. A monument across the ditch marks a Confederate burial site.
10. ARTILLERY PLATFORM: A 32-pounder siege gun fired from this position. When Confederate artillery opened fire from Shepherd's Mountain, this gun dismounted an enemy piece with its first shot, forcing the enemy gunners to withdraw farther down the gap.

11. ARTILLERY PLATFORM: A 3-inch rifle fired from this position. Beyond the gap between Pilot Knob Mountain on the left and Shepherd's Mountain on the right is Ironton, where the fighting began on Sept.26. Artillery fire from the fort kept the Confederates from advancing through the gap.
12. ARTILLERY PLATFORM: A 3-inch rifle fired from this position. The most determined Confederate attacks were made along this wall.
13. ARTILLERY PLATFORM: A 32-pounder siege gun fired from this position. During the third and final Confederate assault, troops of Cabell's Arkansas Brigade jumped into the ditch at this angle but were driven out with hand grenades.
14. GATE AND DRAWBRIDGE (no longer existing): At this point, a drawbridge crossed the ditch to provide access to the fort. During the battle, a Confederate shot cut the drawbridge rope, making it impossible to raise. Twenty Federal volunteers barricaded the gateway and defended it during the assaults.
15. ARTILLERY PLATFORM: A 32-pounder siege gun fired from this position. It was put out of action when an overhead charge caused it to recoil off the platform. The gun was crewed by black citizens from Pilot Knob and Ironton. During the first assault, two guns of Battery H that had been firing from near the town tried to retreat into the fort but were abandoned in front of this angle when several of their horses were killed. The gunners shot the remaining horses to prevent the guns from being captured.
16. POWDER MAGAZINE: This crater is all that remains of the fort's magazine. After the evacuation in the early hours of Sept. 28, a party of volunteers lit the fuse to some 20 tons of ammunition stored under fifteen feet of earth. Amazingly, the Confederates did not investigate the blast until after daylight, by which time the Federals had escaped to the north.

The Ragged Assembly

The expedition was to consist of three mounted divisions, each named after its commanding General. The First Division was to be led by James Fagan, an Arkansas politician with a meager military background; the second was to be headed by John Marmaduke, a West Point graduate; And the third by Joe Shelby, a brilliant and tough cavalry officer.

The three divisions were to move through northeast Arkansas to the Missouri border, where they would split into three columns and advance 20 miles apart on a quick dash into Saint Louis.

Both Fagan's and Marmaduke's divisions were severely undermanned. On his way north through Arkansas, Shelby was ordered to round up as many deserters as possible from behind the Union lines. Shelby rejoined Price's army near the Missouri border, bringing with him more than 3000 deserters at gunpoint to serve in the great campaign. Deserters eventually were to make up nearly a third of Price's force.

Prices entourage was a ragged lot, to say the least. Most of the troops were clothed in tattered rags, and several thousand were barefoot. Most had no canteens, cartridge boxes, or other military issue; Instead, they carried water in jugs and stuffed cartridges in their shirts and pockets.

Tents and blankets were absent. Arms consisted of an endless variety and caliber of rifles and muskets, making ammunition supply in the field nearly impossible. By the time Price reached Missouri, nearly a fourth of his army were without arms.

On September 19th, 1864, Price was ready. A 12,000-man mounted army of military men, misfits, and regulars crossed the border into Missouri.

The Fateful March

Prices advance into Missouri was made by three columns spaced 10 to 20 miles apart. By September 24th, two of the columns had converged on Fredericktown to prepare for a thrust into Saint Louis. Marmaduke joined Price two days later, traveling a longer route.

Meanwhile, in Saint Louis, General William Rosecrans, commander of the Department of Missouri, was getting desperate. Early in September, Rosecrans had received reports that Price was advancing toward Missouri with a major force.

Constant appeals for reinforcements brought only a handful of infantry to defend the city. By late September, a small Garrison of 6000 men was all that stood between the greatest city West of the Mississippi and Price's invading horde.

In Fredericktown, Price and his three division commanders debated whether to assault the federal entrenchment at Pilot Knob on the way to Saint Louis. Having met only token resistance thus far, Shelby wanted to move directly to Saint Louis, which he believed could be taken in a day.

The others felt it would be a tactical mistake to leave an armed Federal Garrison unmolested to the rear as the Confederate column moved north.

On September 26th, the die was cast. Shelby was ordered to move northwest to Irondale and destroy the Saint Louis and Iron Mountain railroad. The rest of Price's army was to prepare for battle.

Without waiting for Marmaduke's entire division to reach Fredericktown, Price ordered Fagan to march north to assault the federal garrison at Pilot Knob.

It was not until the night of September 24 that the Union's General Rosecrans was informed Price's force had crossed into Missouri. As Pilot Knob was his only fortification in south central Missouri, Rosecrans sent St. Louis district commander General Thomas Ewing and a detachment of the 14th Iowa infantry to the area by train.

By noon on September 26, Ewing had reached the hexagonal earthworks known as Fort Davidson.

The fort lay on the floor of a valley surrounded on three sides by commanding hills. It was situated so that the enemy infantry would have to cross hundreds of yards in the open to reach its formidable walls. The fort, however, would be vulnerable to any artillery which could be placed on top of the encircling hills. Ewing had about 1,000 men with which to defend the position.

On the afternoon of September 26, Ewing sent two companies of infantry through Ironton to patrol the roads leading to Fredericktown. No sooner had they reached the "Shut-Ins" gap outside Ironton then they ran head-on into Fagan's advance brigades.

Fagan's Arkansas troops quickly drove the Union patrol back into Ironton, where brisk rifle and cannon fire left scars which still can be seen today on the Iron County Courthouse. Ewing immediately reinforced his patrol with a detachment of the 14th Iowa.

The accurate punishing volleys of the veteran 14th Iowa sent Fagan's untested troops into a near panic, forcing them to retreat to the "Shut-Ins" gap. Repeated attacks by the Confederate advance, however, slowly pushed the Union Skirmishers back into Ironton, where nightfall and a heavy rainstorm brought the engagement to an end.

At dawn of September 27, Fagan's dismounted cavalry, now reinforced by Marmaduke's, hurled themselves at the Union line fronting on the courthouse, forcing a withdrawal to the gap between Pilot Knob and Shepherd Mountains. When the small Union force came within sight of the fort, Ewing ordered the 14th Iowa to a spur of Shepherd Mountain and his dismounted cavalry to the side of Pilot Knob, opening the gap to the federal artillery in the fort.

Heavy skirmishing in the gap resulted in numerous Confederate losses without appreciable gain. Eventually, the desperate Union patrol was overwhelmed and forced to shoot its way back to the rifle pits, which extended from the walls of the fort.

The Silence

Fagan's and Marmaduke's divisions, which already had suffered more than 200 casualties in the first evening and morning of fighting, now swarmed over the encircling hills and into the Ironton gap.

Ewing found himself completely bottled in the fort with no avenue of escape. At a meeting in the gap, Price determined that his big guns would be placed on top of Shepherd Mountain. He then sent an emissary, Colonel Lachlan MacLean, to the fort to ask for a Union surrender.

Hot headed Maclean, a veteran of the Kansas border war, was a personal enemy of Ewing. When the Union general refused to surrender, Maclean returned to Price and urged a frontal assault on the fort, claiming there was no time to bring up all of the Confederate artillery and place it on the mountain.

Price soon became convinced that placing the big guns on the mountain would be no easy task.

Price was now determined to try a frontal assault. For nearly an hour, a hush fell over the peaceful valley- the silence before the storm. Among the heavy brush and timber on the mountains, the Confederate commanders were forming their brigades for battle.

In the fort, Ewing ordered his cannons run down from maximum elevation and trained across the flat. Their load was to be canister rounds, each filled with hundreds of half-inch lead balls. Because all the riflemen could not take their places to file from the walls, details were assembled to tear cartridges, load rounds, and pass up the guns as they were needed.

At the foot of the encircling mountains, 9,000 Confederates crouched down and waited.

The Storm

At two o'clock, the silence was broken. Confederate cannons in the gap opened on the earthen fort. Soon, waves of dismounted southern cavalry poured into the open.

The troops, formed in long columns three ranks deep, slowly moved toward the fort. Inside the walled enclosure, the riflemen were ordered to hold their fire, and the Union artillery was opened on the advancing Confederate lines.

At short range across the flat, the big guns could not miss. Dense clouds of smoke blanketed the fort and rose in columns hundreds of feet high.

The surrounding Confederate mass continued its ill-fated advance. Now aware that the rifle pits could not be held, Union soldiers poured into the fort. The Confederate horde was now only 500 yards from the walls when Union riflemen were ordered to fire.

With spent rifles being passed down and loaded ones handed up, the 300 rifles along the top of the walls spewed forth lead as if from machine guns. Smoke from the heavy fire obliterated the Confederate lines.

At 200 yards, the southern brigades unleashed their first volley and broke into a crazed running charge. The Union Gunners could see only the charging legs of the gun barrels. At 30 yards, Price's troops finally broke and slowly started to fall back.

Spurred on by their gallant officers, the terrified Southerners re-formed their lines and surged ahead. Again, they hesitated and their officers turned them about. This third charge saw some men actually charge into a dry moat that surrounded the fort. The Union gunners, with artillery shells fused as grenades, leaned over the walls and tossed them into the huddled Confederate soldiers.

The blood and confusion now were too much to bear. Just a few yards from the fort, Price's soldiers finally turned and ran. As the soldiers streamed away from the fort and the smoke had a chance to clear, the incredible carnage became apparent.

For 500 yards on the three sides of the fort that were attacked, the ground was covered with dead and wounded men. In the short few minutes that had just passed, one of the bloodiest clashes of the civil war had taken place.

The Great Escape

The black rainy night which settled in the Arcadia valley saw every shelter from Ironton filled with Confederate wounded. Price sent messages north toward the Union lines to beg for medical assistance.

The Tools of War in 1864

Fort Davidson was defended by an impressive concentration of rifled and smoothbore artillery:

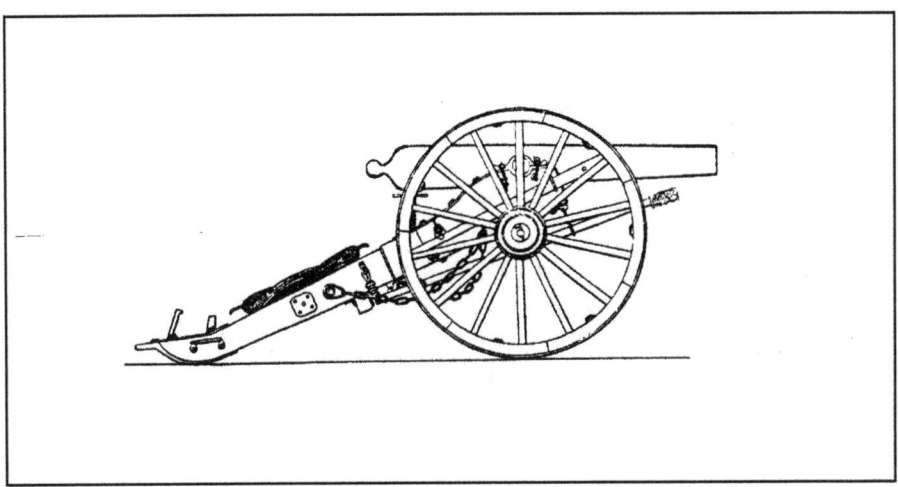

SIX 3-INCH ABOVE ORDNANCE RIFLES – Were manned by Battery H, 2nd Missouri Light Artillery – two fired from outside the fort until just before the first assault, then were abandoned but not captured. Effective range: 2,700 yards with solid shot, shell, or spherical case (shrapnel) and 300 yards with canister (.96-inch iron balls).

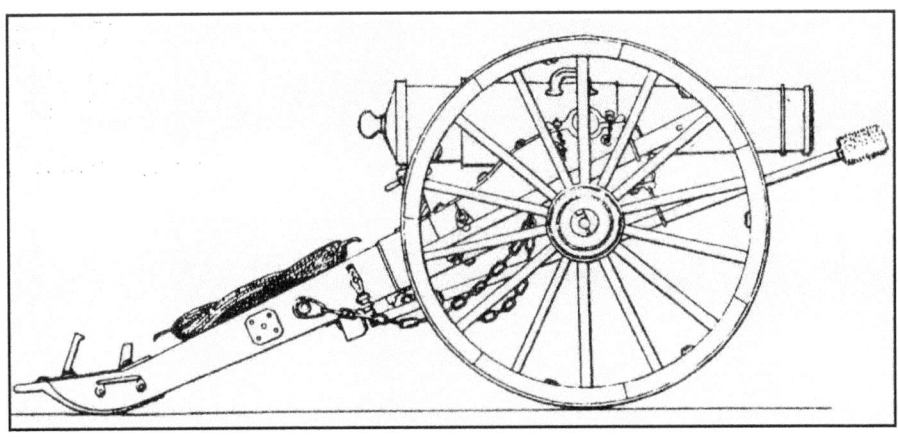

THREE 24-POUDER HOWITZERS Previous Page – Were manned by Company G, 1st Missouri State Militia Infantry. Effective Range: 1,200 yards with shell or spherical case (shrapnel) and 600 yards with canister (48 1.35-inch iron balls).

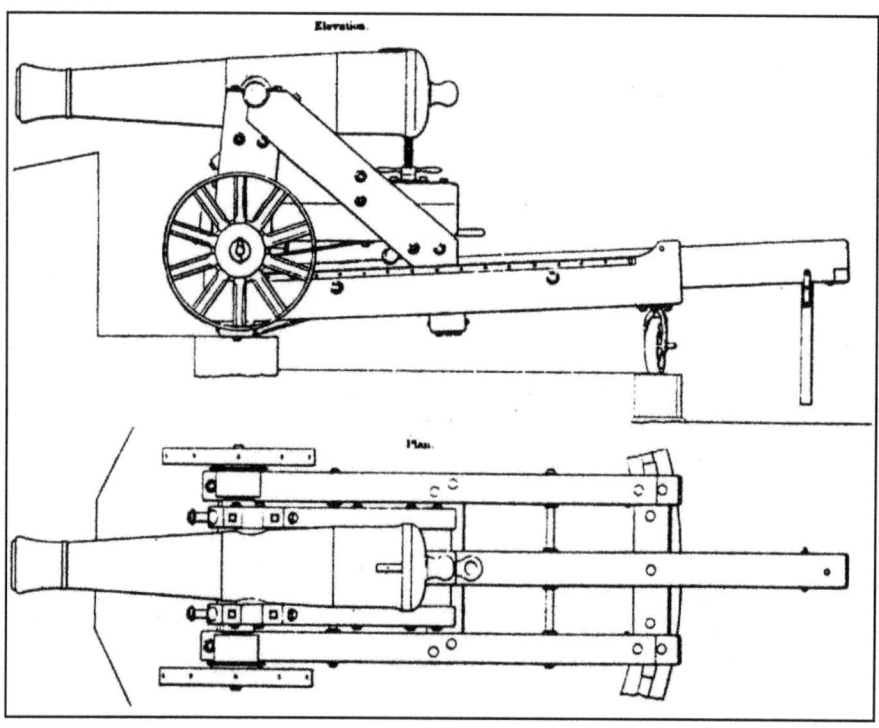

FOUR 32-POUNDER SIEGE GUNS Above – Were manned by Company G, 1st Missouri State MILITIA INFANTRY, and a company of local black citizens. Effective Range: 1,700 yards with solid shot, shell, or spherical case (shrapnel) and 600 yards with canister (27 2-inch iron balls).

His entire command lay in a pitiful state of confusion. Most companies were scattered, and only a few posted centuries or maintained any semblance of military discipline.

Inside Fort Davidson, General Ewing was deciding on his next move. He correctly surmised that the new morning would dawn with Price's artillery perched on top of Shepherd Mountain, rendering the Fort untenable.

Near midnight, Ewing hit upon a daring plan; he would attempt to slip his troops out of the fort and through Confederate lines.

At midnight, Ewing muffled the wheels of the six field guns, with the 14th Iowa at the head, marching the column silently out of the fort. The weary Union defenders moved north along the road to Potosi and miraculously marched unchallenged right through the loose Confederate lines. In a few hours, Ewing was miles away from the fort.

At two o'clock in the morning, a squad left behind in the fort blew up the powder magazine in the center of the earthen enclosure. Confederates roused by the blast thought the explosion was an accident.

At dawn, Price's dwindling army went after the escaping Federals. Although Ewing ran headlong into Shelby, he was able to successfully fight his way to a strong union fortification in Rolla. Marmaduke and Shelby wasted their three days on the futile pursuit.

With his best assault troops lost and two of his divisions in disarray, Price knew that an attack on the now-reinforced city of Saint Louis was out of the question.

To salvage something from the ill-fated campaign, Price decided to turn northwest and capture Missouri's capital for the confederacy. But the weak wasted at pilot knob and the initial crushing defeat had cost him dearly.

Price found that Jefferson City, too, had been reinforced, and he fought only a brief, half-hearted skirmish before marching to final defeat at the Battle of Westport less than a month later.

Chapter 5
PATTERSON CEMETERY

In the area east of this monument are the unmarked graves of both Union and Confederate soldiers. They died during the two fierce battles for the possession of Patterson and Fort Benton: the battles were fought in 1863 and 1864. The latter was the turning point of the war, which ended in 1865. It is fitting that this monument be located in the shadow of Fort Benton and in the cemetery. It is logical that many of these men died in the vicinity where this monument stands. This was a war that did not have to be fought: over 600,000 men died before it was over. This monument is erected in honor of these soldiers and also as a reminder to future generations to realize the horrors of war and to do their utmost to prevent history from repeating itself.

Monument erected by the American Legion Post 281, The Patterson Cemetery Association and the Wayne County Historical Society.

Drone shot from above Fort Benton to the cemetery below, at times, in the shadows.

The Battles at Patterson and Fort Benton

Chapter 6
ARTIFACTS FROM THE FARM - THE BAYONET

It was not surprising to see a bayonet among the items plowed up on my Grandpa's farm. This one is from a Union Springfield rifle.

The Battles at Patterson and Fort Benton

Consider the following from Bruce Catton's book, "The Army of the Potomac Trilogy" -

"Yet it was these ineffective old smoothbores on which all established combat tactics and theories were based. This is why the virtues of the bayonet figured so largely in the talk of professional soldiers of that era. Up until then, the foot soldier was actually a spear carrier in disguise; the bayonet was the decisive weapon, and an infantry charge was just the old Macedonian phalanx in modern dress – a compact mass of men projecting steel points ahead of them, striving to get to close quarters where they could either impale their opponents or force them to run away. All offensive infantry tactics were designed to enable a commander to throw that compact, steel-tipped mass against an enemy line of battle.

But with the rifled musket, it just didn't work that way anymore. The compact mass could be torn to shreds before it got in close. The advancing line came under killing fire four or five times as far as used to be the case. As one student of Civil War casualties remarked: "There was a limit of punishments beyond which endurance would not go, and the old Springfield rifle was capable of inflicting it." Like the machine gun in 1914, there was a weapon that upset all the theories. The natural result was that actual hand-to-hand work with the bayonet was a great rarity in the Civil War, for all the fine talk of grand bayonet charges to be found in the general's memoirs. The bayonet was still carried, and it was still a threat, but very few men ever used it. Of some 245,000 wounds treated by surgeons in Union hospitals, fewer than a thousand had been made with bayonets. One reason, of course, maybe that when a man did get bayoneted, he usually died on the spot; nevertheless, the figure is significant.

The Confederate General John B. Gordon, who got into about as much truly desperate fighting as any man on either side, wrote after the war: "I may say that very few bayonets of any kind were actually used in battle, as far as my observations extended. One line of the other usually gave way under the galling fire of small arms, grape and canister before the bayonet could be brought into requisition. The brisling points and glitter of the bayonets were fearful to look upon, as they were leveled in front of a charging line, but they were rarely reddened with blood." In several private soldiers' memoirs, one finds the remark that the bayonet was really most useful as a candlestick: its point could be jabbed into the ground easily, and its socket was just the right size to hold a candle." (Catton 1951: 195,196)

Smoothbore Molds

One can visualize the camped soldiers gathered around the campfire pouring the "metal to the mold" in preparation for the next days action. It was a multi step process of sorts – melt the lead in a fire, pour the molten lead into a mold, beat off the excess lead that did not fit, let it cool off a bit and then open the mold and drop the piping hot bullet from the mold into the water so it can cool off.

The Battles at Patterson and Fort Benton

"Indiana Memory" put it this way – During the Civil War there were nearly 1,000 different types of bullet in use on both sides. Soldiers often made their own bullets or recycled old or used ammunition by melting it down and remolding it in a similar way to how musket balls were made.

Being made of lead and the effects of lead not totally understood, one can only imagine what lead analysis of the bodies of soldiers on both sides would have revealed. As you can see above, various sizes of smoothbore balls simply called for different size molds.

But Civil War expert Scott House does not agree. He says, "If you are on the move, it is far easier to carry bullets already made than it is to carry raw lead and a heating dish & mold. More likely is that there was a shed or workshop or smith shop where this was done. Requires lots of heat, like charcoal. I remember Jeff Thompson's men latched on to some lead, but moving much of that was impractical."

Bruce Catton notes in his book "The Army of the Potomac Trilogy" –

"Compared with the modern Garand, the rifle was laughable; but compared with the smoothbore which had been the standard weapon in all previous wars, it was terrific.

Early in the Civil War, before the government got rifled muskets into mass production, may regiments were equipped with the old smoothbores, which fired a round ball or, sometimes, a cartridge containing one round ball and three buckshot: the "buck and ball" of army legend. Regiments which had to use such muskets were degusted with them. Extreme range was about 250 yards, and accuracy was almost nil at any range. As one on the backwoodsmen from Wisconsin remarked, it took a fairly steady hand to hit a barn door at fifth paces. At very close range, of course, they were quite effective, especial when firing "buck and ball," which gave a scatter gun effect. These primitive smoothbores were discarded as fast as new weapons were produced, and by the fall of 1862 few regiments on either side carried them." (Catton 1951: 195-195)

The Minie' Ball

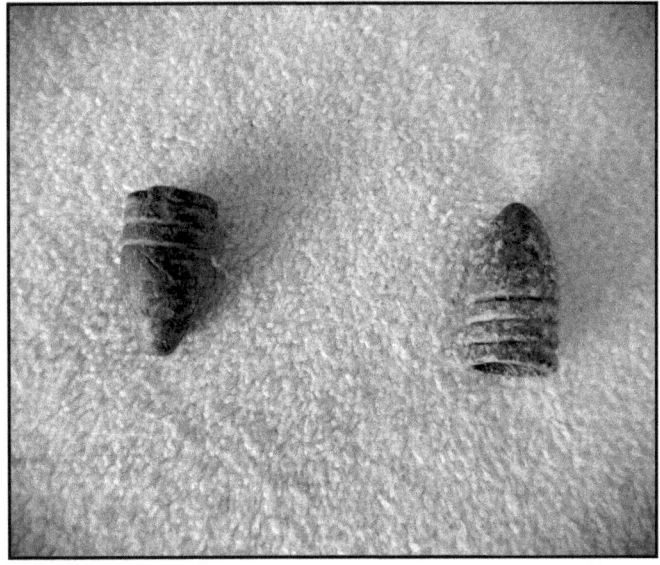

The **Minié ball**, or **Minie ball**, is a type of hollow-based bullet designed by Claude-Étienne Minié, inventor of the French Minié rifle, for muzzle-loading rifled muskets. It was invented in 1846 and came to prominence during the Crimean War and the American Civil War, where it was found to inflict significantly more serious wounds than earlier round musket balls. Both the American Springfield Model 1861 and the British Pattern 1853 Enfield rifled muskets, the most common weapons used during the American Civil War, used the Minié ball.

Rifling, the addition of spiral grooves inside a gun barrel, imparts a stabilizing spin to a projectile for better external ballistics, greatly increasing the effective range and accuracy of the gun. Before the introduction of the Minié ball, balls had to be rammed down the barrel, sometimes with a mallet, because gunpowder residue would foul a rifled bore after a relatively small number of shots, requiring frequent cleaning of the gun. The development of the Minié ball was significant because it was the first projectile type that could be made with a loose enough fit to easily slide down the barrel of a rifled long gun yet maintain good accuracy during firing due to obturation by expansion of the bullet›s base when fired.

Bruce Catton put it this way – "The basic, all-important weapon, of course, was the infantry musket, and the standard of the war was the rifled Springfield. This was a muzzle-loader with an involved procedure for loading. Drill on the target range began with the command, "Load in nine times, load!" (The "nine times" meant that nine separate and distinct operations were involved in loading a piece; recruits were trained to do it "by the numbers.") The cartridge was a paper cylinder encasing a soft lead bullet and a change of powder. The soldier bit off one end of the paper, poured the powder down the barrel, rammed the bullet down with his ramrod, cocked the heavy hammer with his thumb, and had a percussion cap on the nipple to ignite the charge when he pulled the trigger. For most rifles, these caps came in long rolls, which were inserted in a spring-and-cogwheel device in the breech, exactly like the rolls for a child's cap pistol today.

This weapon has long since been a museum piece, but the big point about it then was that it was rifled and had a bullet, which took the riffling properly. The bullet was the Minie, named for the French captain who had invented the bullet. Minnies to all soldiers were a conical slug of lead about an inch long, with a hollow base that expanded when the rifle was fired and prevented leakage of the powder gases. It would kill at half a mile or more, although it was not very accurate at anything like that distance. Its effective range was from 200 to 250 yards – effective range meaning the distance at which a definite line of battle could count on hitting often enough to break up an attack by relatively equal numbers. A good man could fire off two shots in a minute." (Catton 1951: 194)

3 Inch Archer Bolt

The Battles at Patterson and Fort Benton

This 3" Archer Bolt, obviously, is the "cream of the crop" in regard to the items plowed up by my Grand Dad, Roy Kinnison, off his farm located SW from Fort Benton. Noteworthy is the intact lead collar or sabot, which indicates the shell was likely never fired. It is commonly referred to as a 3" Archer Bolt but, in fact, is slightly less in diameter than 3" to fit in a 3" barrel. It is also known as the 3" Rebel in some usage.

The design is credited to Dr. Robert Archer, who was a partner of Joseph Anderson, the Confederate superintendent of the Tredegar Foundry in Richmond, Virginia. It was constructed of cast iron and features a pointed nose that tapered toward the base, part of this obstructed by the sabot. The lead sabot was designed to pick up the rifling.

It had no charge inside and operated much like a large bullet. With no charge, it was often described as ineffective and was often discarded in the field by artillery units. After 1862, it was seldom recovered from the field, making them, in a strange twist, rarer and more valuable.

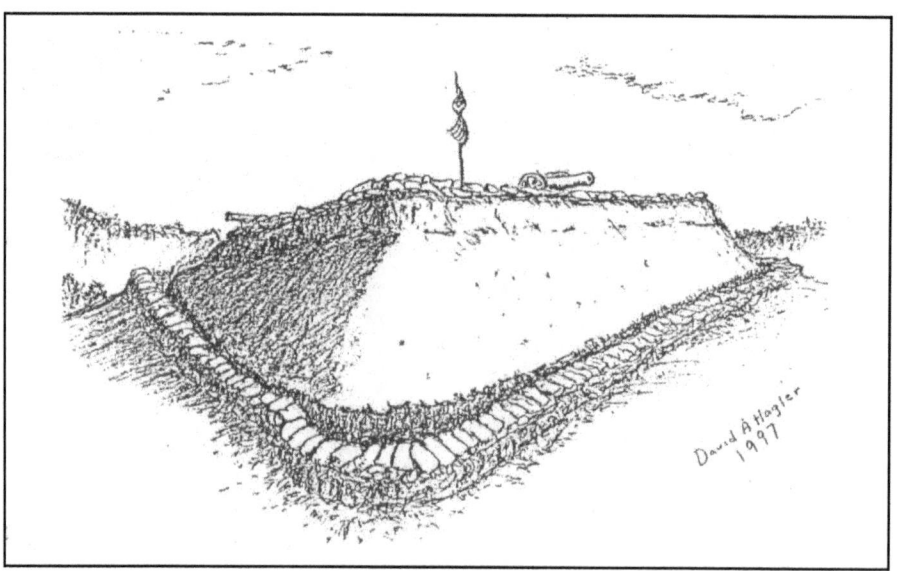

Chapter Seven
TOPOGRAPHY AND DRONE SHOTS

Fort Benton

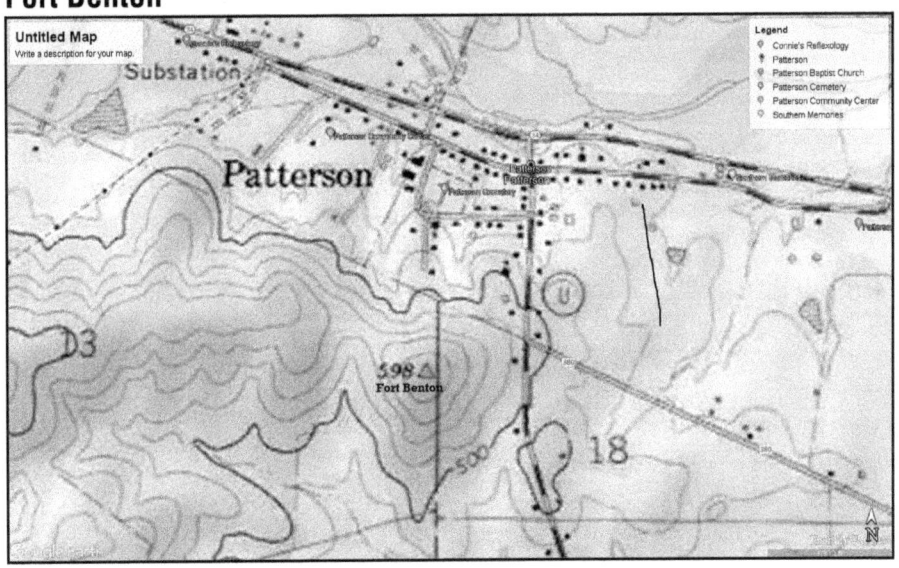

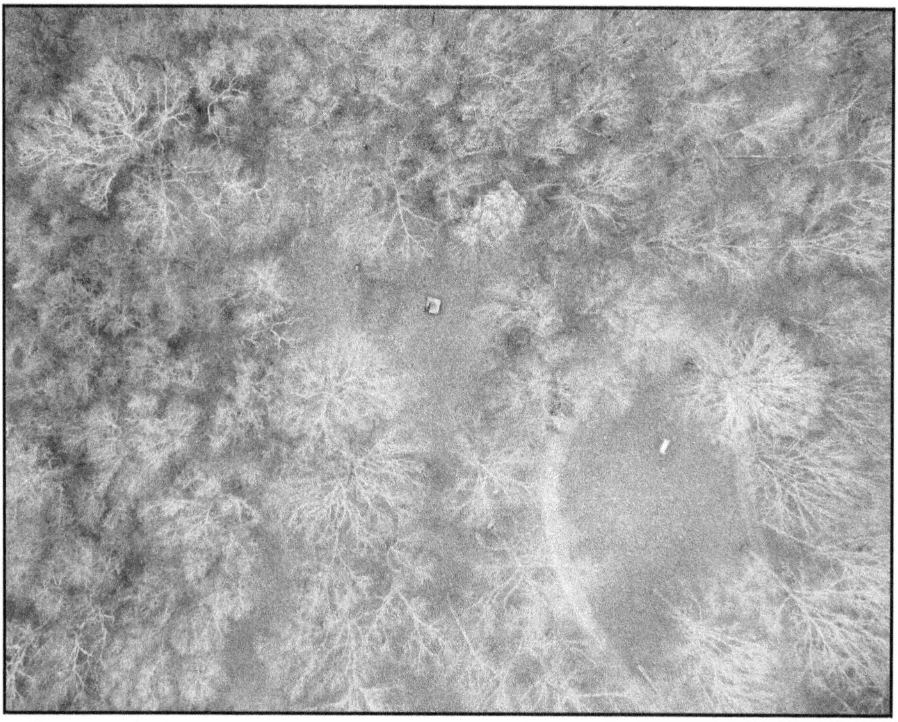

The Battles at Patterson and Fort Benton

Stoney Battery

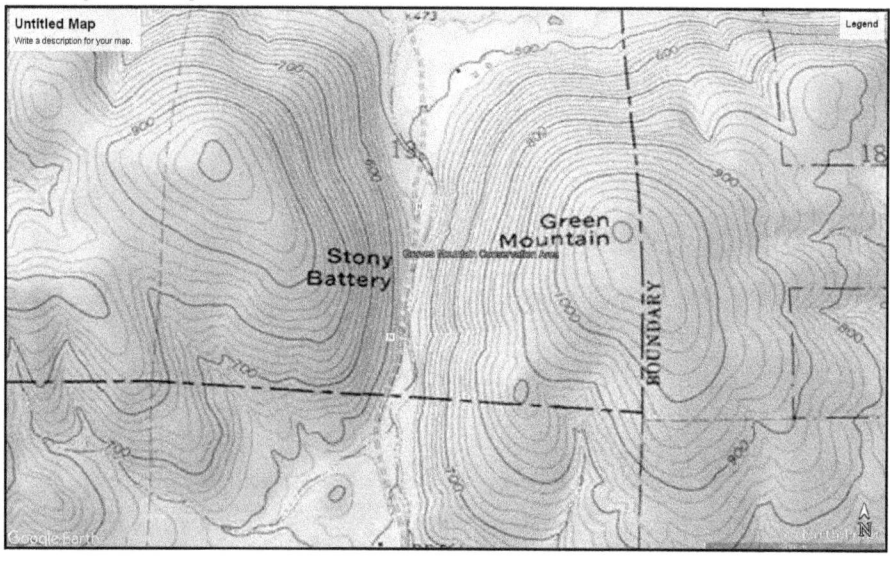

Bob Frakes

Chapter Eight
"EPILOGUE - THE FARM"

To only discuss my Grand Parents farm (from now on "the farm") in only Civil War terms would be incomplete, I think. It certainly is on my top ten list of "growing up factors," and I am sure my cousins feel the same way. It was/is located a quarter mile east on Wayne County 332 from Rings Creek Church.

The People
Note: Much of this information was supplied by my Uncle Ed Kinnison, My Aunt Pat (Kinnison) Young, and my Cousin Charlotte (Bottorff) Branz

The Battles at Patterson and Fort Benton

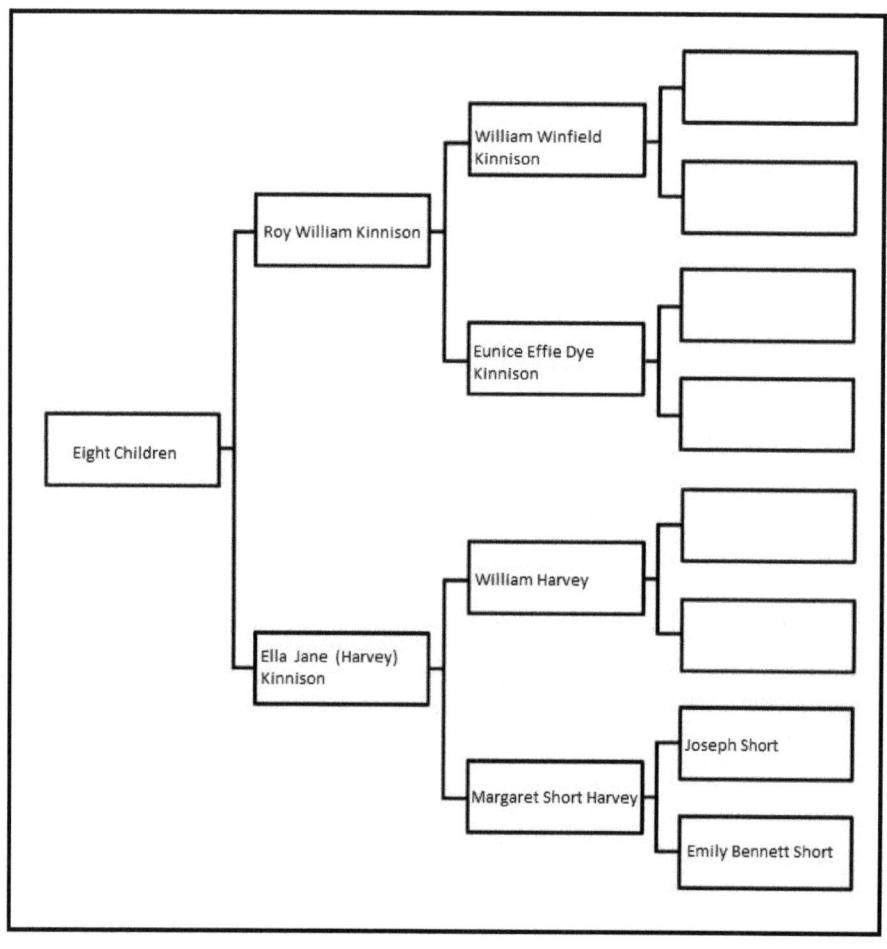

Pictured are my great-grandparents, William Winfield Kinnison and Eunice Dye Kinnison. William was born in Berlin Cross Roads, Ohio and Eunice in Marietta, Ohio. Both passed away in Patterson, Missouri. My Grandpa Roy is in the back, and his sister Mayme Helton is on his right. Lizzie Cook is to his left and may be his sister, but that is not noted on the back of the picture.

Roy William Kinnison (Grandpa Roy to me) – Roy was born April 7th, 1896, in Stronghurst, Illinois. Stronghurst is across the Mississippi from Burlington, Iowa. Burlington remains a Kinnison stronghold, with relatives living there and the final resting place of others. My Mom always mentioned Ohio as the location where many Kinnisons came from moving to Stronghurst and then on to Patterson, Missouri. It seems Roy's Dad came from Berlin Cross Roads, and his Dad from the same. Continuing back, you find Cacapon Springs, Ohio, and Chester County, Pennsylvania before that. Roy's Great, Great, Great, Great Grandfather may have come from England in 1680. The distant Grandmothers and great-grandmothers seem to be more obscure. Roy passed away on August 1st, 1966, and is buried at Patterson Cemetery.

Ella Jane (Harvey) Kinnison (big Grandma to me so as not to be confused with my Dad's Mom, who was a smaller-statured Little Grandma) – Ella was born September 29th, 1895, in Brunot, Missouri, just north of Patterson. On Ella's side, the Dye name goes back generations, where it turns to Duyts from Denmark. Ella passed away September 12th, 1973, and is also buried in Patterson Cemetery. They were married on November 14th, 1917.

Roy and Ella had eight children, four boys and four girls. All were born at Patterson -

Roy William Kinnison Jr. – Married Phyllis Baker/ Children: Patricia Ann Kinnison and Shirley Kay Kinnison.

Meryle Kinnison – Married to William Bottorff, and William Bean/ Children: Charlotte Bottorff, Sharon Bottorff, Donald "Butch" Bottorff, Barbara Bean, Marsha Kay Bean, and William Arlon Bean.

Ella Jane (Pat) Kinnison – Married Kenneth Young/ Children: Janet Lynn Young and Carol Jean Young.

Marjorie Kinnison – Married Harold Frakes/Children: James Frakes and Robert Frakes

Robert Kinnison – Married Mary Lou Gibbs and Betty Eland/ Children: Robert Frank Kinnison, Ronald Kinnison, Robert Gene Kinnison and Roy William Kinnison.

Edward Kinnison – Married Shirley Ann Todd /Children: Sherry Ann Kinnison and Douglas Edward Kinnison.

James Kinnison – Married Charlene Ivester/Children: Charles Kinnison, Rodney Kinnison, Peggy Kinnison, Kirk Kinnison, James Kinnison, Amy Kinnison, Andrew Kinnison and Jay Kinnison.

Mary Celine Kinnison – no children

Pictured up front: Ed Kinnison, Jim Kinnison, and Bob Kinnison. Row 2: William Winfield Kinnison, Ella Kinnison, Meryle Kinnison, Margie Kinnison & Pat Kinnison. Back Row: Roy William Kinnison Jr. & Roy William Kinnison Sr. Celine is on Ella's lap.

Bob Frakes

Here, you will find a picture of most but not all of the cousins. Front Row: Janet Young, Sherry Kinnison, Jean Young, Arlene Bean, Billy Bean and Jim Frakes. Row Two: Marsha Kay Bean, Donald "Butch" Bottorff, Bob Frakes, Charlotte Bottorff, Barbara Sue Bean and Sharon Bottorff.

The Battles at Patterson and Fort Benton

Here is a family story/legend that has been told and retold to me by many. You see my cousin Peggy standing next to a rock. The inscription says:

> In Memory of
> David Kennison
> The Last Survivor of the
> Boston Tea Party
> Who Died in Chicago, February 24, 1852
> Aged 115 Years, 3 Months, 17 Days and is Buried
> Near This Spot. This Stone is Erected
> By The Sons of the Revolution, The Sons of the American Revolution,
> And the Daughters of the American Revolution.

The story is the stone sits in Lincoln Park in Chicago. My Uncle Ed told me historically, the name "Kinnison" was at times spelled "Kennison."

The House

County 332 ran west from Rings Creek Church and jogged south in front of the house and then back west again. The house sat off the road and had a turnaround for the mail delivery at the road. The driveway ran back east to a garage that always had a seldom-driven older model parked in it. A smokehouse sat next to the garage.

The house had a cellar under it and two horizontal barrels next to the cellar door on the south side. They held fuel oil or kerosene, I guess. As you entered the south door, the one most often used, you entered a small eating area, and the kitchen was off to the right. Straight ahead, you stepped up to a living room area that also served as a dining room for the big eats. To the east was Celine's room, and to the north was the bedroom where Grandpa and Grandma slept.

The Battles at Patterson and Fort Benton

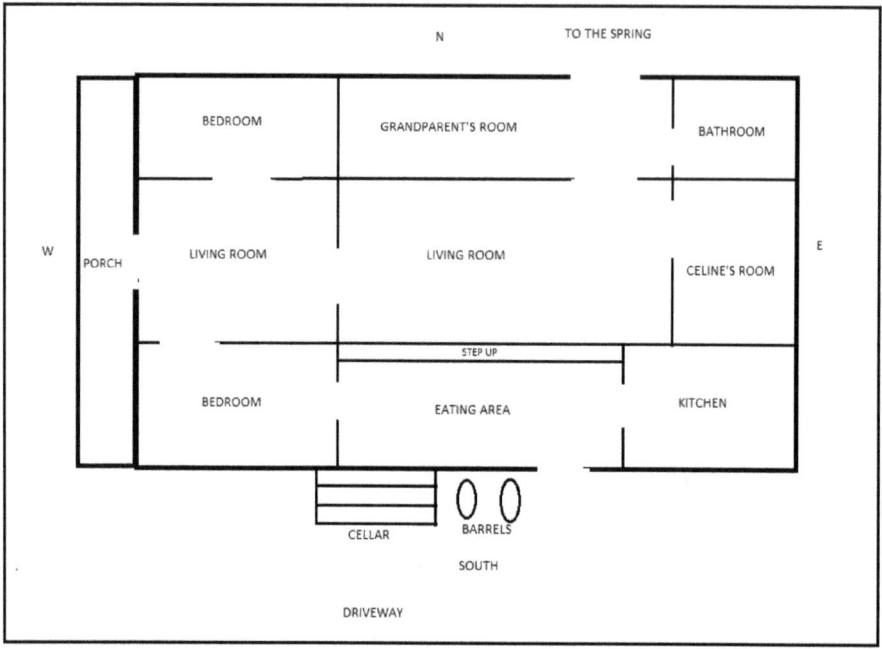

To the west of the big room was a smaller living room with bedrooms to the north and south of that. On occasion, there were relatives everywhere as the farm also served as summer camp and people were sleeping in every room. I recall one time Dad had driven us down and was going to drive back the next morning. As he recalled, every time a car rolled in, everyone had to get up and talk for an hour. At about two in the morning, he threw in the towel and, just got in his car and drove home.

It could be the site of high drama. My cousin Janet came in crying one day, followed by Butch et al. Grandma, the enforcer, conducted the investigation. It seems Butch had shoot her in the butt with a BB gun. Grandma interrogated Burch. He fessed up but claimed Janet had fired first. Grandma asked Janet if that was correct, to which she replied. "Yes, **but I missed!**" All guns were confiscated. I did not witness the BB gun shot out as the older cousins would run off and leave me in the dust now and then on the pretense I was a pest. Imagine that!

One of the high tides each day was breakfast with Grandma's homemade chocolate rolls. Nobody could ever duplicate them; maybe it was the fresh whole milk. Water for the house came from –

The Spring

The farm had several springs. One was located just north of the house. It ran kind of SE and turned south to flow into Rings Creek. It had a culvert inserted horizontally, and you could look down into the culvert and see sand at the bottom with bubbles rising to the surface.

It didn't seem that deep but never ran dry even in drought. When others ran dry, neighbors were always welcome to come and fill up, my cousin Butch told me. Sounds like Grandpa and Grandma. I was told more than once while pestering the older cousins that it was "50 feet deep," and they were going to throw me in. I remember the smell of peppermint and the sight of watermelons bobbing. The cool water kept them cool and ready to eat.

We tried various projects to dam the spring ditch, but the water always won out. Butch and Jim built this "cabin" where the spring ditch joined the creek. Maybe I spent one night? Maybe Mom stayed with us? I do remember the burned fried taters the next morning as "the best I ever ate."

The Orchard

South of the house was a sour apple orchard, brooding house, hen house, gas tank, and some old abandoned farm equipment. I was told during high water, Rings Creek would flow right through the orchard although I never witnessed that. In addition to tree climbing and apple eating, it was always fun to sit on the old equipment and pretend. Being old enough to collect the eggs was a notable point in life. Here is Grandpa on his vintage Ford tractor and bush hog in front of the apple trees. The chicken coop is in the background.

Granary, Shop, and Barn

Running SE from the garage was a series of buildings. A woodlot sat just beyond the gate. There was a "chicken neck cutting" block where chickens began their journey to the dinner table. As a youngster, watching them run around and jump sans a head was pretty perplexing.

The granary was next to the woodlot. You could crawl up and check out the bins of grain and corn. There was a crank device – put the ear of corn in one end and crank. The cob came out one place, and the shelled came the other. Surprisingly, I saw a few rats and mice. I guess the army of cats was there for that – more on that later.

The shop was next. Lots of "stuff" stuffed in there! I always enjoyed turning the forge/blower and watching the dust and debris blow out the center. There was an outhouse there, complete with a Sears & Roebuck catalog. An old "Rio" car sat in the weeds to the east and then the barn.

The story goes Grandpa, and the boys built it with hardwood they cut themselves. I could never figure out how they drove the nails in. Still standing last time I was by. Grandpas' mules Barney and Red called it home. That is also where the milking was done. The cats would line up for Butch to bend an utter and hit them in the face. It had lofts on each side, a central area, and a lift on the west end. On some visits, there was enough hay in the middle to jump out of the loft and land. It was great fun. As I recall, there was another abandoned barn east of the house, but it was grown up all around and I never got very close to it.

Rings Creek (Ring's Creek)

The creek was the site of much fun! There were always some patched tractor tire tubes to use. One hour after lunch (so as to not get the cramps) it was off to the swimming hole being used for that summer.

The creek came out of Mountain (View) Lake, home of the Mountain View Fishing and Hunting Club of St. Louis. The lake was two miles west of the farm. It was spring-fed all the way to the St. Francois River to the east. My cousin Butch had usually picked out a good place and built a stone dam to create a "swimmin hole." It usually created one place deep enough to dive off of, with a leaning tree often adding to the thrill. Here is a picture from a year hole and you can see the stone dam downstream. This one was several hundred yards below where the spring ditch ran in.

Here, we have made harbors for the boats Dad made. Pictured are my brother Jim, Cousin Butch and me. We put firecrackers in drilled holed, lit them, and pushed the boats off to "fire away." I was young and always puzzled at how fireworks could be legal here and not at home? I didn't 'understand different states and different laws yet. The Uncles could put on quite a 4th of July show.

On the south side of the creek below the orchard were "cuts" made into the hill in the war to extract iron ore. And one the far reaches were three stones about the size of basketballs that marked, so the story goes, a pioneer family that got into a dispute with some Native Americans.

What was the prettiest rock you could find? How many times could you skip one? It was good fishing when the creek was fuller. It was good, simple fun.

Snakes Alive!

"Snake safety" was always in the back of our minds. Grandpa Roy was bitten by a Copperhead. Mom was with him as a young girl as they were returning from the creek. The Doctor came out a prescribed the tried and true – soak it in kerosene. His leg swelled up double, Mom said, and he lost circulation and had nerve damage. Grandpa limped for the rest of his life.

There were tales of rattlers bailed up in the hay bales. I recall the last few years I went to the farm, the barn was not being used, and we were told to stay out of the barn as it had become "snakey." Didn't have to tell me twice.

Maybe you remember, I sure do, the gas station in Patterson that sat where the road in from the creek met Rt. 34. There was a stretched-out rattlesnake skin over the door that looked six feet long to a youngster.

Animal Farm

There were so many animals it was hard to tell where to begin. There were horses and mules (this picture is labeled "horse laugh"). There were cows and pigs. I used to throw the skinniest hog an ear of corn. There were chickens and ducks. There was "Ducky Do," a duck from our backyard that was transplanted to the farm when it became clear having a duck in the backyard in town would not work. There were always dogs and cats.

I remember in summer a cat with only one eye. It had lost one in a fight with a rat, was the story. The dogs used to recline in the cool dirt under the oil barrels. There was often one with a big golf ball-sized knot under the chin – snake bite. Grandma would set a dish of clabbered milk out as that was supposed to be a cure.

Then there was the beagle who would take her name from Patterson itself – Patterson Pinky. She had been named Pinky from birth because of her pink nose. She showed promise from the start, as she had an exceptional nose. But, she would not open. You can run the mouth out of them, but little can be done to get them to bark. Dad became disgusted with her and, on one summer trip, took her to the farm and left her. On a later trip, in one of those destiny moments, he literally had one leg in the car to go home and heard a dog running on the hill across the road. He asked Grandpa what dog that was and he answered maybe Pinky as she had been barking some. Dad

The Battles at Patterson and Fort Benton

took off on foot, brought her home and she quickly won enough licensed trials (four in one year) for her AKC Field Championship, Dad's first.

Her clinching win involved a recently oiled road. The oil was not totally wet but not dry either. The judges watched from horseback as the rabbit crossed the oiled road. The pack stopped except Pinky, who took the rabbit right across the oil. The head judge declared in a drawl, "Well, boys, you can pick 'em up!" You may be able to pick up the oil on her feet and leg before Dad gets her cleaned up.

NAMESAKE BEAGLE OF PATTERSON A FIELD CHAMPION

Harold D. Frakes, Secretary of the King City Beagle Club, Mt. Vernon, Illinois, reports that a beagle, namesake of Patterson in Wayne county, has won her field championship.

Frakes briefs the "wins" of Patterson Pinky, named after the town of Patterson.

Field Champion Patterson completed her Championship Sunday, September 29, by winning at the Three Rivers Beagle Club, Paducah, Ky., with 33 entries. This was her second win within two weeks and her fourth Licensed Win in less than one year.

The other wins were as follows: October 5, 1956, first at Hawkeye Beagle Club, Columbia, Mo., 33 entries; October 27, 1956, first at Okaw Beagle Club, St. Elmo, Ill., 32 entries; September 27, 1957, first at Pere Marquette Beagle Club, Beaver Dam, Ill., 26 entries.

Mr. Frakes, in reporting the successful wins of Patterson Pinky, says: "When Pinky was a puppy I took her down to my father-in-laws farm, Roy Kinnison of Patterson, there she learned to run a rabbit," "I brought her back to Mt. Vernon and trained her for field trial work."

"In addition to being a Field Champion," said Mr. Frakes, "she is also a wonderful mother, whelping 11 puppies in April of 1956 and 10 in April of 1957."

Rings Creek School

When it came to walking a mile to and from school, the Kinnison kids did, 1.15 miles to be exact. Rings Creek School was located about a mile west of the house. It was one room, eight grades, maybe? I know Mom graduated from Patterson High School. She was barely 16 when she graduated, as it was the practice back then to just push you ahead a grade if you could handle it. She would work as a seamstress, get her LPN license later, rise to the head of nursing at a nursing home, and finally do corporate work for the company that owned that facility and others.

Mom recalled one day when a boy took a swing at the teacher. The teacher just sent a note home, and the boy's Dad came to school and took him out by the collar. He returned the next day "all lined out," and that was the end of that.

She also remarked how much you could learn as the teacher worked with the upper grades. My Aunt Celine had Down's Syndrome, but Mom noted all she learned in the one-room setting.

Here is a picture from a few years ago. I don't know if the building is still up or not.

Mom graduated from Patterson High School.

The Battles at Patterson and Fort Benton

Mom, Marjorie Kinnison, is fourth from the left in the front row.

"The Farm" from Fort Benton

The farm sat roughly within the oval on the map. The house and barn are below the arrows. This was all south and west of Fort Benton. Where exactly the war artifacts were plowed up is not known.

PICTURE NOTES & CREDITS

All pictures are from my collection or the public domain except "The Fort" side view sketch – David Hagler
"The Fort" overhead sketch – Piedmont City web page done originally by Joe Huett
Pictures of the 150 anniversary of the clash at Stoney Battery by Patti House
Route of the ill-starred invasion and list of battles – Civil War Centennial Commission pamphlet
Pilot Knob sketch, Fort Davidson diagram, and tools of war – Fort Davidson/Pilot Knob State Historic Site

BIBLIOGRAPHY/REFERENCES

Catton, Bruce (1951) Mr. Lincoln's Army 194 – 196.
Catton, Bruce (1952) Glory Road - General
Catton, Bruce (1953) A Stillness at Appomattox – General
Fort Benton Fell Twice – written by David Hagler for the July 1^{st,} 1997 River Hills Traveler Magazine section "Traveling into History." Adapted with the author's and magazine's permission
Missouri – A Divided State in a Divided Nation - Adapted from a pamphlet by the Civil War Centennial Commission
Pilot Knob Battle – Adapted with the permission of the River Hills Traveler Magazine – September 2017.

NOTES

The focus of this book was basic information on Fort Benton and the Battles of Patterson, Missouri, in the Civil War. However, other topics, useful in "setting the table" and following up were addressed.

For the Civil War in Missouri, I adapted pieces from "The Civil War in Missouri," which was published by the Civil War Centennial Commission of Missouri. This pamphlet offered an excellent basic look at that topic. It was not copyrighted, but it was used in my effort to "not reinvent the wheel." The pamphlet acknowledged the contributions of David Brown, St. Louis Post Dispatch, Dr. Richard S. Brownlee - Director of the State Historical Society of Missouri, Missouri Historical Society, and Missouri State Park Board.

Likewise, Fort Davidson & the Battle of Pilot Knob was not my focus but needed some attention. The section on that in the book was adapted from a Missouri Department of Natural Resources piece that appeared in the River Hills Traveler Magazine and was used with their permission.

The piece on Fort Benton was done many years ago by David Haggler and published in the River Hills Traveler Magazine. It was used and adapted with the permission of both.

REFERENCES/SUGGESTED READING

Brownlee, Richard S. Gray Ghosts of the Confederacy. LSU Press, Baton Rouge 1958.
Britton, Wiley. The Civil War on the Border. New York, London, 1891.
Carr, Lucien. Missouri, A Bone of Contention. Boston and New York,1888.
Castel, Albert. William Clarke Cantrill: His Life and Times. New York, 1962.
Connelley, William E. Quantrill, and the Border Wars. Cedar Rapids, Iowa 1910.
Edwards, John N. Noted Guerrillas, or The Warfare of the Border. St. Louis,1877.
Monaghan, Jay. Civil War on the Western Border, 1854-1865. Boston, Toronto,1955.
O'Flaherty, Daniel. General Joe Shelby. Chapel Hill, N.C., 1954
Rea, Ralph R. Sterling Price, Pioneer Press, Little Rock, Arkansas, 1959
Ryle, Walter H. Missouri Union or Secession, Nashville, 1931
Schofield, John M. Forty-Six Years in the Army. New York, 1897
Smith, Edward C. The Borderland in the Civil War. New York, 1927
U. S. War Department. The War of the Rebellion: A Compilation of the Official Records

INDEX

Symbols

3/5 Compromise 4

A

Abolitionists vii,1,8
Alien & Sedition Acts vii, 5
Anderson, Bloody Bill 26,28,30
Archer Bolt vii,65
Articles of Confederation 3

B

Battle of Westport 31
Battles of Patterson xi
Bayonet vii,59
Bean, Barbara 74,75
Bean, Arlene 75
Bean, Billy 75
Bean, Marsha Kay 74,75
Bean, William Arlon 74
Benton, William Plummer 34
Blair, Francis 17
Blair, Francis P. Jr. 16
Bleeding Kansas 11
Boonville 19

Bottorff, Charlotte 74,75
Bottorff, Donald "Butch" 74,75
Bottorff, Sharon 74,75
Branz, Charlotte 71
Brooks, Preston 11
Brown, John 12

C

Camp Jackson 17
Clement, Archie 27,28
Columbus 21
Colonies vii,2
Compromise of 1850 9
Continental Congress 3
Cook, Lizzie 72
Crawford, Jeptha 27

D

Douglas, Stephen 10
Dred Scott Decision vii, 12

E

Elkhorn Tavern 23

F

Fagan, James F. 30,45
Fort Benton vii,33,67
Fort Davidson 35,40,45
Fort Sumter 16
Frakes, James 74,75
Frakes, Robert 74,75
Fredericktown 49
Fremont, General 21
Frost, D.M. 17

G

Gadsden Purchase 7
Gag rule 1
Garrison, William Lloyd 8
Grant, Gen. Ulysses S. 21
Greenville 35
Guerrilla Wars vii, 23

H

Halleck, Henry Wager 22
Harney, William Selby 16

I

Indentured Servents 3

J

Jackson, Andrew 7
Jackson, Claiborne Fox 16
Jackson, Governor 18
James, Frank 27
James, Jesse Woodson 27

Jayhawkers 27

K

Kansas-Nebraska Act vii,10
Kennison, David 74
King Cotton 2
Kinnison, Amy 74
Kinnison, Andrew 74
Kinnison, Betty 74
Kinnison, Charles 74
Kinnison, Charlene 74
Kinnison, Ed 74
Kinnison, Ella Jane 72,74
Kinnison, Eunice Dye 72
Kinnison, James 74
Kinnison, Jay 74
Kinnison, Kirk 74
Kinnison, Marjorie 74
Kinnison, Mary Lou 74
Kinnison, Meryle 74
Kinnison, Pat 74
Kinnison, Patricia Ann 74
Kinnison, Peggy 74,76
Kinnison, Robert 74
Kinnison, Robert Frank 74
Kinnison, Robert Gene 74
Kinnison, Rodney 74
Kinnison, Ronald 74
Kinnison, Roy William 74
Kinnison, Roy William Jr. 72,74
Kinnison, Sherry Ann 74,75
Kinnison, Shirley Kay 74

Kinnison, William Winfield 72

L
Leeper, Captain W.T. 38
Lexington 21
Liberator 8
Lincoln Park 76
Louisiana Purchase vii, 5

M
Marmaduke, John S. 29,36,38,48
McElroy, Captain Robert 40
Mexican Cession 7
Mexican War vii,7,8
Middle Colonies 2
Minie' Ball vii,63
Missouri Compromise vii,1,6,12

N
Northwest Ordinance 3

O
Old Pap 19,29

P
Patterson 33,35
Patterson Cemetery vii,57
Patterson Pinky 85,86
Pea Ridge 31
Pike, Albert 22
Pilot Knob vii, 35
Pilot Knob Mountain 50

Polk, Leonidas 21
Popular Sovereignty 11
Price, General Sterling 18,19,29

Q
Quantrill, William C. 26,28,30

R
Reynolds, Thomas C. 29,30
Rings Creek vii,79,80,82,87
Rolfe, John 3
Rosecrans, General W.S. 30

S
Scott, Dred 12
Samuel Slater 4
Smart, Colonel Edwin 36
Smoothbore Molds vii,61
Stoney Battery vii,68
Stowe, Harriet Beecher 10
Sumner, Charles 11

T
Todd, George 26,28
Trans-Mississippi Dept. of the Confederacy 22,24
Turner, Nat 8,9

U
Uncle Tom's Cabin vii, 10
Underground Railroad 8

V

Virginia and Kentucky Resolutions 5

W

War of 1812 vii, 5
Wayne County 35
Whitney, Eli 6
Wide Awakes 16,17
Wilson's Creek 20

Y

Young, Carol Jean 74,75
Youngers, Cole and Jim 27
Young, Janet Lynn 74,75
Young, Pat 71

www.ingramcontent.com/pod-product-compliance
Lightning Source LLC
Chambersburg PA
CBHW050733010526
44107CB00010B/840